Perspectives on Morphological Organization

Empirical Approaches to Linguistic Theory

Series Editor

Brian D. Joseph (*The Ohio State University,* USA)

VOLUME 10

The titles published in this series are listed at *brill.com/ealt*

Perspectives on Morphological Organization

Data and Analyses

Edited by

Ferenc Kiefer
James P. Blevins
Huba Bartos

BRILL

LEIDEN | BOSTON

The Library of Congress Cataloging-in-Publication Data is available online at http://catalog.loc.gov

Typeface for the Latin, Greek, and Cyrillic scripts: "Brill". See and download: brill.com/brill-typeface.

ISSN 2210-6243
ISBN 978-90-04-34291-0 (hardback)
ISBN 978-90-04-34293-4 (e-book)

This book is printed on acid-free paper and produced in a sustainable manner.

Contents

Editors' Introduction

Ferenc Kiefer, James P. Blevins, and Huba Bartos

Aims and Scope

This book contains a selection of papers presented at the 16th International Morphology Meeting held in Budapest, from May 31 through June 3, 2014. The first set of papers are selected from the general session of the conference, while the second part contains papers presented at the workshop on information-theoretic approaches to morphology. The papers in the two parts of the volume discuss related morphological phenomena but offer radically different explanations.

The first part contains papers on aspects of the organization of morphological paradigms, which have not received sufficient attention thus far. The topics include the organization of inflectional paradigms in non-Indo-Eurorpean languages, a discussion of the question as to whether paradigm complexity is stem or affix based and a novel look at inflectional markings of morphosyntactic properties. The latter includes a new look at the types of exponent relations, extends the notion of periphrasis to inflectional classes, discusses the role of polyfunctionality in inflectional morphology, the non-canonical behavior of subjects and objects and their case markings, the relationship between inflectional class productivity and morphological change, as well as paradigm leveling in Russian and its theoretical consequences.

The second part explores the application of information theory and discriminative learning models to the analysis of morphological systems. These papers fall within a long tradition of information-theoretic analyses, which runs from estimations of statistical properties of English in Shannon's initial papers (Shannon 1948, 1951) to the development of the entropy-based psycholinguistic processing models described in Milin et al. (2009). The current papers extend this tradition by using information theory to address a number of basic questions regarding the organization of morphological systems. One set of issues concerns the communicative dynamics which govern the interaction between regular and irregular patterns or which determine the functional load of gender oppositions and other grammatical contrasts. A second set of issues relate to the formal reconstruction of traditional morphological notions in terms that facilitate the assessment of sources of complexity and the evaluation of competing divisions of grammatical labor. A third group arises in connection with the interface to syntactic processes—particularly with respect to the load

 | DOI 10.1163/9789004342934_002

imposed by the online interpretation of morphological constructions. The papers in this second part were collected following a workshop on Information Theory in Morphology at the 16th International Morphology Meeting. One outcome of the workshop was a clearer understanding of the close relation between notions of 'morphological information' based on 'uncertainty' and 'uncertainty reduction' and the error-driven structure of discriminative learning models. In effect, measures of system entropy can be seen as providing a global estimation of the local uncertainty associated with the transitions of a processing model that incorporates a discriminative learning network. A second outcome was a recognition that the convergence of research strands described above has now produced a general model of language structure and use that 'scales up' to stretches of the speech stream larger than the word and 'scales down' to stretches smaller than the word.

Part 1

In his paper entitled *Polyfunctionality and the Variety of Inflectional Exponence Relations*, Gregory Stump (University of Kentucky) offers a novel explanation of the polyfunctionality of inflectional affixes. In inferential-realizational theories of inflection, the only grammatically significant relation that exists between inflectional markings and morphosyntactic properties is that of exponence; but variable affix ordering and morphotactic conditioning reveal that affixes participate in three sorts of exponence relations—inherent exponence (an exponence relation that is independent of the position of the affix in a word form's morphotactics), positional exponence (an exponence relation that is dependent on an affix's morphotactic position), and conflated exponence (an exponence relation expressed by a combination of affixes). The latter two relations contribute to the polyfunctionality of the inflectional morphology of a language, as the author demonstrates with evidence from the Swahili system of noun-class concords. Furthermore, the distinction between inherent and positional exponence suggests that rules of exponence actually consist of two independent parts: exponence declarations specify an affix's inherent exponence; sequencing rules specify an affix's linear ordering and its positional exponence. Thus, one source of polyfunctionality is the existence of formatives with fixed inherent exponence that acquire different sorts of additional content from different sequencing rules. Another way in which an affix can acquire additional content is through affix conflation, a process that combines the form and content of two affixes to produce a complex affix with its own distinctive function and distribution. Stump develops an analytic framework that makes it possible to say that each Swahili noun-class concord has constant bit of content (the noun class of which it is an inherent exponent) but may express additional content according to its position (expressing subject concord in one affix position, object concord in another) or to its conflation with another affix (e.g. the default relative affix, with which a noun-class concord combines to form an affix expressing agreement with a relative verb form's relativized argument). This framework provides additional motivation for the assumption that a language's morphology constitutes an autonomous, irreducible grammatical component.

The second paper (*Verb Morphology and Conjugation Classes in Dunan*) is devoted to the description of the verb morphology of Dunan, a highly endangered language spoken on Yonaguni Island. At the very beginning of their paper, the authors, Thomas Pellard (CNRS-CRLAO, INALCO-CRLAO) and Masahiro Yamada (Ritsumeikan University, Kyoto) point out that Japanese and other Japonic languages have a relatively simple and transparent verb morphology, with few non-canonical phenomena. They generally exhibit a highly agglutinative structure with little morphophonology, few conjugation classes, limited stem allomorphy, and very few irregular verbs. This explains why morpheme-based constructive approaches to morphology have been popular for the description of these languages. In contrast, the verb morphology of Dunan, a highly endangered Ryukyuan language spoken by around 500 people on Yonaguni Island, is much more complex and departs in several interesting ways from the simpler system exhibited by its relatives. In particular, the existence in Dunan of a rich system of conjugation classes, of a rather high degree of allomorphy, and of several non-canonical phenomena, is noteworthy. The analysis presented in the paper is based on original data gathered during fieldwork that fill important gaps in previous descriptions. It also provides a further argument for the observation that smaller languages with few speakers tend to have a richer verb morphology (more inflectional classes, more allomorphy) than large languages with many speakers. It is argued that neither morpheme-based morphophonology, nor a root/stem and paradigm based morphology can do justice to the Dunan data. The authors propose an account in terms of a modified word and paradigm model in which it is assumed that forms do not exist in isolation but only as part of a structured system of interdependencies and implicational relations.

Inflection classes have been presented as a prototypical instance of a morphomic category (Aronoff 1994), lacking any identifiable morphosyntactic function. Traditionally, inflection classes are defined in terms of affix homonymy. However, recent work has focused on the role of stem alternation in defining inflectional complexity. For example, Bonami & Boyé (2002) have argued that the inflectional complexity of French verb classes can be most economically defined in terms of stem sets connected by 'dependency relations' defined by an inheritance hierarchy. Noura Ramli (University of Essex) argues in her paper *Inflection Classes in Transitional Libyan Arabic* (*TLA*) that, like French, TLA realizes morphosyntactic features by affixation, and TLA verbs fall into inflection classes defined in terms of stem alternations (the inflectional exponents themselves show no allomorphy even for irregular verbs). The morphomic stem allomorphy of the perfect tense series forms bears on a wider question which has figured prominently in morphological debates in recent years, especially in Semitic linguistics: should we analyze Arabic morphology in terms of an autosegmental morpheme-based approach or in terms of a stem-based approach? The morphomic patterning revealed in TLA conjugation poses serious empirical problems for any morpheme-based account. Therefore, following the Bonami & Boyé (2002) approach to French, the author defines an inheritance hierarchy for TLA morphomic verb stems and shows how this effectively defines a set of inflection classes in the absence of affixal allomorphy.

Ekaterina Georgieva's (Research Institute for Linguistics, Hungarian Academy of Sciences) paper entitled *Person Agreement on Converbs in Udmurt* is a revealing account of a morphosyntactic change affecting converbs in Udmurt. The change involves reanalysis of the earlier pronominal agreement on converbs into grammatical agreement with pro-drop. The author shows that person agreement on converbs has mixed properties: with respect to referentiality it remains pronominal, at the same time it may also allow subject-doubling of the originally pronominal suffixes. For speakers who use subject-doubling person agreement on converbs, pronominal agreement has been turned into grammatical agreement with pro-drop. This change fits nicely into the typological observations on how agreement arises. At the same time, the paper also sheds light on some hitherto unexplored aspects of Udmurt morphosyntax.

Typically, language spread goes hand in hand with the simplification of its morphology. Varvara Magomedova (SUNY, Stony Brook) and Natalia Slioussar (HSE, Moscow and St. Petersburg State University) show how this tendency can be observed in Russian morphology. In their paper *Paradigm Leveling: The Decay of Consonant Alternations in Russian* they show how paradign leveling affects old consonant alternations. In standard Russian, such alternations are not infrequent. The paper focuses primarily on comparatives and some verb forms. It is shown, among other things, that if a verb has a standard form with alternations, speakers will use the verb without difficulty. On the other hand, problems arise with novel verbs that do not belong to literary (standard) Russian and with verbs that exhibit paradigm gaps. In the case of adjectives, alternation loss occurs frequently with compound adjectives whose second member can be used as an independent word with alternation. The relevant data are taken from internet texts which reflect the everyday usage of the language.

Part 2

In the first paper in this part, *The Zipfian Paradigm Cell Filling Problem*, James P. Blevins (University of Cambridge), Petar Milin (University of Sheffield), and Michael Ramscar (University of Tübingen) offer a novel perspective on the interaction of regular patterns and irregular formations in a morphological system. The chapter focuses on two competing communicative pressures that can reach different states of equilibrium in different languages. The first is a discriminative pressure towards maximal differentiation of forms and the messages they express, which enhances variation between expressions. In most languages, this discriminative pressure is countered by a predictive pressure, which favors regular patterns of form and distribution that facilitate the prediction of unencountered forms. While it has long been known that corpora provide only a partial coverage of the forms of a language, this chapter presents evidence that the shortfall is far greater than previously appreciated, and that the coverage of form variation remains sparse in corpora regardless of size. The chapter also suggests how the interaction between discrimination and regularity gives rise to systems of form contrasts and meaning contrasts in which lexical neighborhoods play a central role. Since most paradigms will be only partially attested, clustering paradigms into neighborhoods provides an analogical base for the deduction of unencountered forms. This perspective leads to the expectation that there will be a correlation between similarity at the level of form within neighbourhoods and co-filled paradigm cells that bootstraps analogical deduction.

The second study, *Paradigm Realization and the Lexicon*, Géraldine Walther (CNRS and Universität Zürich) falls, within a complementary information-theoretic perspective that uses minimum description length to measure morphological complexity. The chapter begins with a systematic reassessment of the role and nature of the realizational lexicon. Although realizational approaches are firmly anchored in a lexicalist conception of grammar, existing realizational approaches tend to focus primarily on the format of rules or their organization, leaving the interface between inflectional paradigms and the lexicon largely implicit. This is especially clear in the literature on paradigmatic irregularity, which contains comparatively little work concerned with the representation of irregularities within inflectional lexical entries. A general reconsideration of the lexical component of realizational approaches reopens basic questions about the division of labor within these approaches. These questions lead in turn to a reappraisal of core notions such as 'inflectional classes,' 'inflectional categories,' 'morphosyntactic features,' and 'inflectional regularity.' This chapter explores these issues by outlining a realizational model that specifies some of the notions that remain implicit in other models—providing, in particular, an explicit definition of realizational lexical entries, and defining the relationship between paradigmatic structures in a language and the properties of individual lexical entries in a language. This model not only highlights the origin and locus of lexeme-specific inflectional irregularities, but also offers precisely formalized definitions of the inflectional notions indicated above.

In the third paper, *The Possible Role of Entropy in Processing Argument Dependencies in Hungarian*, Csaba Pléh (Central European University), István Fekete (Carl von Ossietzky Universität, Oldenburg), and Dániel Varga (Rényi Institute of Mathematics, Hungarian Academy of Sciences) explore the use of entropy measures to predict response latencies in processing ambiguous case constructions. The paper outlines an information-theoretic analysis of the processing load associated with oblique verbal dependents in Hungarian which are, as a class, systematically ambiguous between argument and modifier interpretations. In place of the verb frames and thematic roles used in most previous analyses, this information-theoretic alternative uses entropy measures computed from the MOKK and MAZSOLA corpora to estimate the uncertainty of continuations in oblique dependent constructions. The experiments reported in this chapter find robust effects of entropy in verb-dependent processing and weaker effects associated with the entropy of frames. The fact that entropy measures account for many of the effects associated with frames in previous studies leads the authors to suggest that differences in entropy could in principle explain the result pattern more parsimoniously than a putative representational-semantic difference between comitative and instrumental constructions.

In the fourth study, *A Functional Theory of Gender Paradigms*, Melody Dye (Indiana University), Petar Milin (University of Sheffield), Richard Futrell (MIT), and Michael Ramscar (University of Tübingen) offer an explanation for the seemingly perplexing organization of gender systems. They argue that the source of perplexity lies in a mischaracterization of the function of gender as taxonomic. Instead, they suggest, noun class serves a discriminatory purpose, and the information processing requirements of discriminatory systems are very different from those of taxonomic systems. A striking result of this perspective is the discovery that the gender system in German, far from being unsystematic and meaningless, not only serves to reduce the entropy of nouns in general, but is more specifically informative about high frequency nouns than low frequency nouns. More generally, masculine and feminine gender classes in a language like German do not reflect a deep underlying taxonomic distinction; rather, items are assigned to different gender classes because assigning them to different gender classes is informative. In a system with only two types of members—male and female—gender classification can appear to be taxonomically interpretable. But in a system with many types of members, such as drinks and days of the week, the same classification appears taxonomically senseless. From a discriminative perspective, the underlying logic is consistent in both cases: Gender serves to redistribute the entropy of the nouns, making them more predictable, on average, in context.

PART 1

CHAPTER 1

Polyfunctionality and the Variety of Inflectional Exponence Relations[1]

Gregory Stump

1 Exponence Relations

In inferential-realizational theories of inflection, the only grammatically significant relation that exists between inflectional markings and morphosyntactic properties is that of exponence (Stump 2001:11). This premise places important restrictions on the analysis of inflected word forms. Consider, for example, the Latin verb form *rēxī* 'I have ruled.' One imaginable analysis for this form is schematized in Figure 1.1. In this analysis, the desinence *-ī* expresses the first person singular indicative and selects a perfect active stem (such as *rēk-s*). Thus, *-ī* is related to morphosyntactic property sets in two ways: it is an exponent of {1 sg ind}, and it is contextually restricted to {perf act} stems. This distinction between properties of content and properties of context is excluded in inferential-realizational theories of inflection, which instead entail the analysis schematized in Figure 1.2. In this analysis, there is a property set {1 sg perf ind act} to be realized, and it is this property set that (in the inflection of the lexeme REGERE 'rule') dictates both the choice of the perfect active stem *rēk-s* and that of the first-person singular perfect indicative active desinence *-ī*.

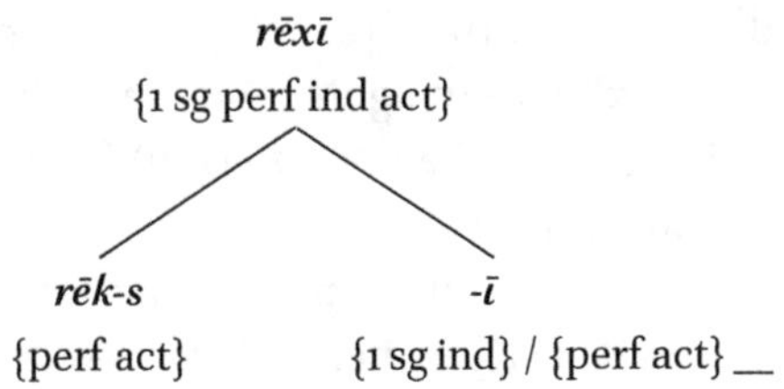

FIGURE 1.1 **rēxī** *'I have ruled.'*

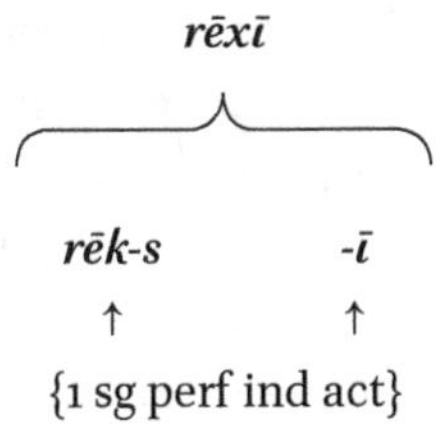

FIGURE 1.2 **rēxī** *'I have ruled.'*

1 An earlier version of this paper was presented at the 16th International Morphology Meeting, held in Budapest in May–June 2014; several of those present gave helpful comments, for which I am grateful. I have also benefited from helpful discussions with Olivier Bonami and Berthold Crysmann.

© KONINKLIJKE BRILL NV, LEIDEN, 2017 | DOI 10.1163/9789004342934_003

Thus, the suffix *-ī* stands in a single relation to a morphosyntactic property set: it is an exponent of {1 sg perf ind act}.

Yet, there are different kinds of exponence. In particular, the phenomena of variable affix ordering and morphotactic conditioning reveal that an affix may participate in three sorts of exponence relations: intrinsic, positional, and conflated. Swahili verb morphology presents instances of all three of these exponence relations (§2); the latter two of these contribute to the polyfunctionality of the language's inflectional morphology. I present an inferential-realizational framework that accounts for all three kinds of exponence, exemplifying its characteristics with a detailed analysis of the Swahili evidence. In particular, I introduce a distinction between intrinsic exponence declarations and sequencing rules and demonstrate how this distinction accounts for intrinsic and positional exponence (§3); I then show how conflated exponence declarations give rise to the phenomenon of conflated exponence (§4). Summarizing the proposed framework, I conclude by highlighting its advantages for the analysis of the Swahili facts and by suggesting a number of wider applications in morphosyntax (§5).

2 Intrinsic, Positional, and Conflated Exponence in Swahili Verb Morphology[2]

Swahili possesses a noun-class system, a kind of gender system in which a noun's gender is reflected not only in the inflection of agreeing constituents but also in its own inflection for number. Thus, a noun is typically associated with two noun classes: its singular form inflects as a member of one class and its plural form, as a member of a distinct class; each class has an inventory of morphological markings used to inflect an agreeing verb, adjective, determiner, or pronoun. I refer to a given gender with a pairing *x*|*y* such that a noun belonging to *x*|*y* exhibits the noun-class prefix *x-* in the singular and the prefix *y-* in the plural; thus, as a member of the ki|vi gender, the noun KITABU 'book' has the singular form *ki-tabu* and the plural form *vi-tabu*.

Verbal concords in Swahili distinguish person, number, and (in the third person) gender, and express both subject and object agreement. Subject agreement and object agreement are marked in different positions in a verb form's prefixal string; but as Table 1.1 shows, the prefix used to express agreement with a subject's person, number, and gender properties is often identical in form to the prefix used to express object agreement for those same properties. For the

2 See Ashton 1944 for extensive discussion and examples of the phenomena described here.

TABLE 1.1 *Swahili verbal concords*

Person		Singular		Plural	
1st		*ni-*		*tu-*	
2nd		(sbj.)	(obj.)	(sbj.)	(obj.)
		u-	*ku-*	*m-*	*wa-*
3rd	Gender				
	m|wa	*a-*	*m-*	*wa-*	
	m|mi	*u-*		*i-*	
	ki|vi	*ki-*		*vi-*	
	ji|ma	*li-*		*ya-*	
	n|n	*i-*		*zi-*	
	u|n	*u-*		*zi-*	

purposes of exemplification, I focus on the verbal concord *vi-*, but the characteristics of *vi-* that are at issue here can be equivalently exemplified with any of the concords in Table 1.1.

The concord *vi-* is an exponent of the property set

{{GEND:ki|vi, PER:3, NUM:plural}};

in what follows, I shall use ρ as an abbreviation for exactly this property set. As examples (1)–(3) show, the concord *vi-* is used in three different ways in Swahili verb morphology. It expresses agreement with a plural subject belonging to the ki|vi gender, as in (1); it expresses agreement with a plural object belonging to this gender, as in (2); and in the inflection of the verb in a relative clause, it combines with the formative *o* (*vi-o* → *vyo*) to express agreement with a plural relativized argument belonging to the ki|vi gender, as in (3). In this third use, it has special morphotactics; thus, in (3), *vi-* appears after the verb stem rather than before it.

(1) Subject agreement in Swahili
Vi-tabu vi-me-anguka.
ρ-book SBJ:ρ-COMPL-fall.down
'The books have fallen down.'

(2) Object agreement in Swahili
U-me-vi-ona *vi-tabu?*
2SG.SBJ-COMPL-OBJ:ρ-see ρ-book
'Have you seen the books?'

(3) Agreement with a relativized argument
vi-tabu *a-vi-taka-vy.o* *Hamisi*
ρ-book SBJ:m|wa.SG-OBJ:ρ-want-REL:ρ Hamisi.m|wa.SG
'the books which Hamisi wants'

We can therefore distinguish three kinds of exponence exhibited by the concord *vi-*. It is an INTRINSIC exponent of ρ: that is, it is intrinsically an exponent of third person, plural number, and the ki|vi gender. It also exhibits POSITIONAL exponence, expressing agreement either with a subject possessing the property set ρ or with an object possessing this property set, depending on its position in a verb form's morphology. Finally, it exhibits what I shall call CONFLATED exponence, joining with the formative *o* to express agreement with a relative verb form's relativized argument. These three kinds of exponence are summarized in Table 1.2.[3]

TABLE 1.2 *The three kinds of exponence relations in which vi- participates*

Kind of exponence	Content expressed by *vi-*	Context
intrinsic	{{ki|vi 3 pl}}	*vi-* intrinsically has this content (though it never has ONLY this content)
positional	{{sbj ki|vi 3 pl}} {{obj ki|vi 3 pl}}	*vi-* has this content in different positions
conflated	{{rel ki|vi 3 pl}}	*vi-* has this content when conflated with *-o*

3 Table 1.2 reflects the assumption that when a verb inflects for agreement with a subject (or object or relativized) argument α, the morphosyntactic properties of α belong to a set {sbj ...} (or {obj ...} or {rel ...}) which itself functions as a member of the verb's morphosyntactic property set; for instance, a verb inflecting for future tense and first-person singular subject

3 Accounting for Intrinsic and Positional Exponence

How is the distinction between intrinsic and positional exponence to be modeled theoretically? Inferential-realizational theories of inflection often fail to distinguish these two sorts of exponence, representing them as concomitant effects of the same rule. For example, the rule of exponence in (4) simultaneously specifies both the content and the morphotactics of the English plural suffix *-s*.[4] The distinction between intrinsic and positional exponence suggests that rules of exponence should actually consist of two independent parts:[5]

- **exponence declarations** specify intrinsic content;
- **sequencing rules** specify an exponent's linear ordering and its positional content under that ordering.

This reconception of rules of exponence affords a straightforward account of the inflectional polyfunctionality exhibited by (1) and (2): in these sentences, a formative with fixed intrinsic content realizes different positional content according to the particular sequencing rule that defines its morphotactics in any given case.

(4)	An English rule of exponence:	X,	N,	{plural}	→	X*s*
				↑		↑
				content expressed by *-s*		morphotactics of *-s*

Thus, consider the analysis of Swahili verb inflection in Table 1.3. The upper half of Table 1.3 lists several intrinsic exponence declarations. Each declaration has the form in (5), which represents an affixal exponent X of property set σ.

agreement has the property set {fut {sbj 1 sg}}. Thus, the property set {{ki|vi 3 pl}} in Table 1.2 is unspecific about whether the properties 'ki|vi,' '3' and 'pl' are properties of subject agreement, object agreement, or agreement with a relativized argument.

4 A rule of exponence in the format in (i) is interpreted as being applicable to the pairing $\langle Z, \sigma \rangle$ of a stem Z with a property set σ only if Z belongs to category C and σ is an extension of τ; in that case, the result of applying this rule fo $\langle Z, \sigma \rangle$ is the pairing $\langle f(Z), \sigma \rangle$. (Concerning the relevant notion of 'extension', see Gazdar et al. (1985:27), Stump (2001:41).).

(i) X, C, $\tau \rightarrow f(X)$.

5 This proposal is reminiscent of the idea, entertained by Stump (1993: 174f), of defining affixal exponence separately from affix linearization, much as relations of immediate dominance and linear precedence are defined separately in GPSG (Gazdar et al. 1985: 46) and its successors.

(5) $\left[\!\!\left[\begin{matrix}\mathrm{X}\\ \sigma\end{matrix}\right]\!\!\right]$ (or ⟦X, σ⟧)

TABLE 1.3 *Swahili verb inflection (Part 1)*

Intrinsic exponence declarations

a. Tense affixes

⟦ *ta*, {fut} ⟧
⟦ *li*, {past} ⟧
⟦ *me*,{completive} ⟧

b. Noun-class concords

⟦*u*, {{sbj 2 sg}}⟧	⟦*tu*, {{m|wa 1 pl}}⟧	⟦*li*, {{ji|ma 3 sg}}⟧
⟦*m*, {{sbj 2 pl}}⟧	⟦*wa*, {{m|wa 3 pl}}⟧	⟦*ya*, {{ji|ma 3 pl}}⟧
⟦*a*, {{sbj m|wa 3 sg}}⟧	⟦*u*, {{m|mi 3 sg}}⟧	⟦*i*, {{n|n 3 sg}}⟧
⟦*ku*, {{obj 2 sg}}⟧	⟦*i*, {{m|mi 3 pl}}⟧	⟦*zi*, {{n|n 3 pl}}⟧
⟦*wa*, {{obj 2 pl}}⟧	⟦*ki*, {{ki|vi 3 sg}}⟧	⟦*u*, {{u|n 3 sg}}⟧
⟦*m*, {{obj m|wa 3 sg}}⟧	⟦*vi*, {{ki|vi 3 pl}}⟧	⟦*zi*, {{u|n 3 pl}}⟧
⟦*ni*, {{m|wa 1 sg}}⟧		

Sequencing rules

Block −1. Object agreement rules : For every noun-class concord ⟦X, {τ}⟧, the Block −1 sequencing rule is Pref(⟦X, {{obj} ∪ τ}⟧).

Block ±2. Tense rules :
For every tense affix ⟦X, σ⟧, the sequencing rule is Pref(⟦X, σ⟧).

Block −3. Subject agreement rules : For every noun-class concord ⟦X, {τ}⟧, the Block −3 sequencing rule is Pref(⟦X, {{sbj} ∪ τ}⟧).

The first of these exponence declarations identify the intrinsic content of three tense affixes; the remainder specify the intrinsic content of the noun-class concords in Table 1.1. These intrinsic exponence declarations show the content that each affix serves to realize in any context in which it appears; but they do not give any morphotactic information (not even indicating whether the declared affixes are prefixal or suffixal).

The ways in which these affixes combine with stems are instead specified by the sequencing rules defined in the lower half of Table 1.3. The rules are organized into blocks that determine the relative sequence of their application; thus, each block of sequencing rules corresponds to a different affix position in the classification represented in Figure 1.3.

	negative prefix *ha-*	subject agreement	tense or negative prefix *si-*	object agreement	verb root	final vowel	relative suffix
Affix positions:	−4	−3	±2	−1		1	±2
Rule blocks:	−4	−3	±2	−1		1	

FIGURE 1.3 *Affix positions in Swahili verb inflection and the corresponding blocks of sequencing rules.*

Block −1 contains the rules realizing object agreement, Block ±2 the rules realizing tense, and Block −3 the rules realizing subject agreement. These sequencing rules all have one or the other of the forms in (6), where $[\![X, \sigma]\!]$ is a declared affix and f is a function in the set of property sets. In either form, the rule is applicable to the pairing $\langle Y, \tau \rangle$ of a stem Y with a property set τ only if τ is an extension of $f(\sigma)$. When one or the other sort of rules in (6) applies to $\langle Y, \tau \rangle$, the result is (a) $\langle XY, \tau \rangle$ or (b) $\langle YX, \tau \rangle$, respectively.

(6) a. $\text{Pref}([\![X, f(\sigma)]\!])$
b. $\text{Suff}([\![X, f(\sigma)]\!])$

The Block ±2 rules are the simplest sequencing rules, since they themselves don't realize any content other than the intrinsic content of the tense markers they affix; that is, the function f in these rules is simply an identity function. The rules in Blocks −1 and −3, by contrast, all realize positional content in addition to the intrinsic content of the exponents they affix; that is, $f(\sigma) \neq \sigma$ in these rules. Thus, given the exponence declaration in (7a), the analysis in Table 1.3 defines two sequencing rules: rule (7b) belongs to Block −1, and its application causes *vi* to serve in position −1 as a positional exponent of object agreement; by contrast, rule (7c) belongs to Block −3, and its application causes *vi* to serve in position −3 as a positional exponent of subject agreement.

(7) a. Noun-class concord $\left[\!\!\left[\begin{matrix} vi \\ \{\{\text{ki}|\text{vi 3 pl}\}\} \end{matrix}\right]\!\!\right]$

b. Block −1 sequencing rule : $\text{Pref}\left(\left[\!\!\left[\begin{matrix} vi \\ \{\{\text{obj ki}|\text{vi 3 pl}\}\} \end{matrix}\right]\!\!\right]\right)$

c. Block −3 sequencing rule : $\text{Pref}\left(\left[\!\!\left[\begin{matrix} vi \\ \{\{\text{sbj ki}|\text{vi 3 pl}\}\} \end{matrix}\right]\!\!\right]\right)$

Most of the Swahili noun-class concords exhibit this sort of positional exponence, which is one of the sources of polyfunctionality in Swahili verb inflection. Nevertheless, there is no sequencing rule in Block −1 for the affixes ⟦*u*, {{sbj 2 sg}}⟧, ⟦*m*, {{sbj 2 pl}}⟧, ⟦*a*, {{sbj m|wa 3 sg}}⟧, since property sets instantiating the pattern {{obj} ∪ {sbj ...}} are always ill-formed. Similarly, there is no sequencing rule in Block −3 for the affixes ⟦*ku*, {{obj 2 sg}}⟧, ⟦*wa*, {{obj 2 pl}}⟧, ⟦*m*, {{obj m|wa 3 sg}}⟧, since property sets instantiating the pattern {{sbj} ∪ {obj ...}} are likewise ill-formed.

This analysis provides a simple and accurate account of the positional exponence of noun-class concords. Consider, for example, the verb form *vi-me-anguka* '(books) have fallen down.' This form realizes the cell ⟨*anguka*, {completive {sbj 3 pl ki|vi}}⟩ in the paradigm of ANGUKA 'fall down.' Abbreviating the property set {completive {sbj 3 pl ki|vi}} as σ, we can represent the realization of ⟨*anguka*, σ⟩ as in Table 1.4. No rule from Block −1 is applicable, since σ doesn't express object agreement. In Block ±2, the sequencing rule for the completive affix *me* applies, realizing the tense specification in σ. In Block −3, the sequencing rule for the noun-class concord *vi* applies, realizing both the intrinsic content of *vi* (its properties of person, number and gender in (7a)) and its positional content as an expression of subject agreement (in (7c)).

TABLE 1.4 *Swahili* vi-me-anguka *'(books) have fallen down'*

Where σ = {completive {sbj 3 pl ki|vi}},

Cell :		⟨*anguka*, σ⟩
Block −1 :	not applicable	↓ ⟨*anguka*, σ⟩
Block ±2 :	Pref(⟦*me*, {completive}⟧) :	↓ ⟨*me-anguka*, σ⟩
Block −3 :	Pref(⟦*vi*, {{sbj 3 pl ki|vi}}⟧) :	↓ ⟨*vi-meanguka*, σ⟩

4 Accounting for Conflated Exponence

The phenomenon of conflated exponence is exhibited by Swahili relative affixes, which appear on relative verb forms (verb forms in relative clauses) and serve to express the gender and number of a relative verb form's relativized argument. The inventory of relative affixes is given in Table 1.5.

The relative affixes expressing the m|wa gender stand apart from the others in the sense that neither is morphologically complex. Although it is used to mark relative verb forms whose relativized argument is a plural member of the m|wa class, the relative affix *o* is in fact the default relative marker; its content is simply {{rel}}. Noun-class concords combine with this default relative affix *o* to produce relative affixes for all genders other than the m|wa gender; as Table 1.6 shows, such combinations involve various phonological modifications of the noun-class concords.

The content of each complex relative affix listed in Table 1.6 is deducible from the content of its parts. For instance, the noun-class concord *vi* (with content {{ki|vi 3 pl}}) combines with the default relative affix *o* (with content {{rel}}) to produce a complex affix *vyo* having the content {{rel ki|vi 3 pl}}. A complex affix arising from this sort of combination embodies the phenomenon of conflated exponence defined in (8) and exemplified in (9). Thus, each of the nonpersonal noun-class concords in Table 1.3 (i.e. those expressing neither first nor second person nor the personal, m|wa gender) combines with the default relative affix ⟦*o*, {{rel}}⟧ to produce the conflations in (10). By virtue of this conflation, the Swahili noun-class concords take on a third function, that of expressing noun-class agreement with a verb's relativized argument. The relative affixes therefore embody a second source of inflectional polyfunctionality in

TABLE 1.5 *Swahili relative affixes*

Gender	Singular	Plural	
m|wa	*ye*	*o*	← basic relative affixes; *o* is the default relative affix
m|mi	*o*	*yo*	} relative affixes arising through conflation
ki|vi	*cho*	*vyo*	
ji|ma	*lo*	*yo*	
n|n	*yo*	*zo*	
u|n	*o*	*zo*	

TABLE 1.6 *Verbal concords and conflated relative affixes*

	Verbal concords		Conflated relative affixes	
Gender	Singular	Plural	Singular	Plural
m|mi	*u-*	*i-*	*o* (← *u-o*)	*yo* (← *i-o*)
ki|vi	*ki-*	*vi-*	*cho* (← *ki-o*)	*vyo* (← *vi-o*)
ji|ma	*li-*	*ya-*	*lo* (← *li-o*)	*yo* (← *ya-o*)
n|n	*i-*	*zi-*	*yo* (← *i-o*)	*zo* (← *zi-o*)
u|n	*u-*	*zi-*	*o* (← *u-o*)	*zo* (← *zi-o*)

the Swahili verb system: an affix takes on an additional function by conflating with another affix.

(8) Conflated exponence

If $\left[\!\left[\begin{smallmatrix} x \\ \sigma \end{smallmatrix}\right]\!\right]$ and $\left[\!\left[\begin{smallmatrix} y \\ \tau \end{smallmatrix}\right]\!\right]$ are affixes, the **conflation** of $\left[\!\left[\begin{smallmatrix} x \\ \sigma \end{smallmatrix}\right]\!\right]$ and $\left[\!\left[\begin{smallmatrix} y \\ \tau \end{smallmatrix}\right]\!\right]$ is $\left[\!\left[\begin{smallmatrix} xy \\ g(\sigma, \tau) \end{smallmatrix}\right]\!\right]$, where $g(\sigma, \tau)$ is the smallest well-formed property set that is an extension of both σ and τ.

(9) $\left[\!\left[\begin{smallmatrix} \textit{vi} \\ \{\{\text{ki|vi 3 pl}\}\} \end{smallmatrix}\right]\!\right]$ and $\left[\!\left[\begin{smallmatrix} \textit{o} \\ \{\{\text{rel}\}\} \end{smallmatrix}\right]\!\right]$ conflate as $\left[\!\left[\begin{smallmatrix} \textit{vyo} \\ \{\{\text{rel ki|vi 3 pl}\}\} \end{smallmatrix}\right]\!\right]$.

(10) Conflated relative affixes

⟦*o*, {{rel m|mi 3 sg}}⟧	⟦*yo*, {{rel ji|ma 3 pl}}⟧
⟦*yo*, {{rel m|mi 3 pl}}⟧	⟦*yo*, {{rel n|n 3 sg}}⟧
⟦*cho*, {{rel ki|vi 3 sg}}⟧	⟦*zo*, {{rel n|n 3 pl}}⟧
⟦*vyo*, {{rel ki|vi 3 pl}}⟧	⟦*o*, {{rel u|n 3 sg}}⟧
⟦*lo*, {{rel ji|ma 3 sg}}⟧	⟦*zo*, {{rel u|n 3 pl}}⟧

The distribution of a conflated relative affix such as *vyo* parallels that of the unconflated relative affix *ye*. In the default case, the relative affixes are suffixal, as seen earlier in (3). But in the presence of a prefix expressing tense or negation (what I shall call a 'TN affix'), the relative affixes are prefixal, as in (11) and (12).

(11) *vitabu* *a-na-vyo-vi-soma*
book.ki|vi.PL SBJ:m|wa.SG -TNS-REL:ki|vi.PL-OBJ:ki|vi.PL-read
Hamisi
Hamisi.m|wa.SG
'the books which Hamisi is reading'

(12) *vitabu* *a-si-vyo-vi-taka*
book.ki|vi.PL SBJ:m|wa.SG-NEG-REL:ki|vi.PL-OBJ:ki|vi.PL-want
Hamisi
Hamisi.m|wa.SG
'the books which Hamisi doesn't want'

This unexpected change in the location of the relative affix justifies the postulation of an affix position ±2 that includes mutually exclusive prefixes and suffixes. Thus, in relative verb forms, the appearance of a relative affix in suffixal position is mutually exclusive with that of a prefixal TN affix; if a relative verb form is marked for tense or negation, relative agreement is achieved by means of a conflated affix—the conflation of a TN affix with a relative affix—and this conflated affix has the prefixal distribution of a tense affix occupying position ±2. For example, the negative affix ⟦*si*, {neg −tense}⟧ and the relative affix ⟦*vyo*, {{rel ki|vi 3 pl}}⟧ conflate as in (13); the sequencing rules for all three of the affixes in (13) are situated in Block ±2, and are therefore mutually exclusive in their application.

(13) Conflated TN affix, e.g.

$$\left[\!\!\left[\begin{array}{c} si \\ \{\text{neg} -\text{tense}\} \end{array}\right]\!\!\right] \text{ and } \left[\!\!\left[\begin{array}{c} vyo \\ \{\{\text{rel ki}|\text{vi 3 pl}\}\} \end{array}\right]\!\!\right] \text{ conflate as}$$

$$\left[\!\!\left[\begin{array}{c} sivyo \\ \{\text{neg} -\text{tense } \{\text{rel ki}|\text{vi 3 pl}\}\} \end{array}\right]\!\!\right].$$

Thus, a simple and precise account of the phenomenon of conflated exponence in Swahili verb inflection may be formulated as in Table 1.7. This analysis is an extension of the analysis of positional exponence sketched in Table 1.3. The intrinsic exponence declarations are supplemented by the addition of the negative affix ⟦*si*, {neg −tense}⟧ (which joins with the tense affixes as members of the class of TN affixes) and the basic relative affixes ⟦*ye*, {{rel m|wa 3 sg}}⟧ and ⟦*o*, {{rel}}⟧. Two conflated exponence declarations are also included: one augments the class of relative affixes with the addition of conflated relative affixes (each the conflation of a noun-class concord with the default relative affix ⟦*o*, {{rel}}⟧), and the other augments the class of TN affixes with conflated TN affixes (each the conflation of a TN affix with a relative affix). Thus, five kinds of affixes are now declared in the analysis in Table 1.7; these are

- noun-class concords, e.g. ⟦*vi*, {{ki|vi 3 pl}}⟧
- relative affixes
  - intrinsic, e.g. ⟦*ye*, {{rel m|wa 3 sg}}⟧
  - conflated (= noun-class concord with relative affix *o*), e.g. ⟦*vyo*, {{rel ki|vi 3 pl}}⟧
- TN affixes
  - intrinsic, e.g. ⟦*si*, {neg −tense}⟧
  - conflated (= TN affix with relative affix), e.g. ⟦*sivyo*, {neg −tense {rel ki|vi 3 pl}}⟧

TABLE 1.7 *Swahili verb inflection (Part 2)*

Intrinsic exponence declarations

a. TN affixes

⟦*ta*, {fut}⟧	⟦*me*, {completive}⟧
⟦*li*, {past}⟧	⟦*si*, {neg −tense}⟧

b. Noun-class concords (as in Table 1.3)

c. Relative affixes

⟦*ye*, {{rel m|wa 3 sg}}⟧
⟦*o*, {{rel}}⟧

Conflated exponence declarations

Conflated relative affixes

The conflation of a noun-class concord and the relative affix ⟦*o*, {{rel}}⟧ ***is a relative affix.***

Conflated TN affixes

The conflation of a TN affix and a relative affix ***is a TN affix.***

Sequencing rules

Block −1. (as in Table 1.3)

Block ±2. TN rules

For every TN affix ⟦X, τ⟧, the sequencing rule is Pref(⟦X, τ⟧).

For every relative affix ⟦X, τ⟧, the sequencing rule is Suff(⟦X, τ⟧).

Block −3. (as in Table 1.3)

The sequencing rules in this analysis are as in Table 1.3, except that Block ±2 is now formulated as containing sequencing rules of two types: rules for prefixing (basic or conflated) TN affixes and rules for suffixing (basic or conflated) relative affixes. Competition between these two sorts of Block ±2 sequencing rules is invariably resolved by Pāṇini's principle. For instance, Block ±2 contains the two sequencing rules in (14); in realizing the cell ⟨*taka*, σ⟩, where σ is as in (15), (14b) overrides (14a) as the narrower rule.

(14) a. $\text{Suff}\left(\left[\!\!\left[\begin{array}{c} \textit{vyo} \\ \{\{\text{rel ki|vi 3 pl}\}\} \end{array}\right]\!\!\right]\right)$

b. $\text{Pref}\left(\left[\!\!\left[\begin{array}{c} \textit{sivyo} \\ \{\text{neg } \{\text{rel ki|vi 3 pl}\}\} \end{array}\right]\!\!\right]\right)$

(15) σ = {neg {sbj 3 sg m|wa} {obj 3 pl ki|vi} {rel 3 pl ki|vi}}

This analysis affords a straightforward account of the phenomenon of conflated exponence. For example, the relative verb form *asivyovitaka* '(books) that s/he doesn't want' realizes the cell ⟨*taka*, σ⟩ in the paradigm of TAKA (where σ is again as in (15)). We can define the realization of ⟨*taka*, σ⟩ as in Table 1.8. In Block −1, the sequencing rule for the plural ki|vi concord applies, realizing both the concord's intrinsic content in (7a) and the positional content of object agreement in (7b). In Block ±2, the sequencing rule for the conflated TN affix *sivyo* applies, realizing the specifications for negation and relative agreement in σ. In Block −3, the sequencing rule for the subject concord *a* applies, realizing both the intrinsic content of *a* (= {{sbj m|wa 3 sg}}) and (redundantly) its positional content as an expression of subject agreement.

TABLE 1.8 *The realization of Swahili* asivyovitaka *'(books) that s/he doesn't want' (σ as in (15))*

Cell:		⟨*taka*, σ⟩
		↓
Block −1 :	$\text{Pref}\left(\left[\!\!\left[\begin{array}{c} \textit{vi} \\ \{\{\text{obj 3 pl ki|vi}\}\} \end{array}\right]\!\!\right]\right)$:	⟨*vi-taka*, σ⟩
		↓
Block ±2 :	$\text{Pref}\left(\left[\!\!\left[\begin{array}{c} \textit{sivyo} \\ \{\text{neg } \{\text{rel 3 pl ki|vi}\}\} \end{array}\right]\!\!\right]\right)$:	⟨*sivyo-vitaka*, σ⟩
		↓
Block −3 :	$\text{Pref}\left(\left[\!\!\left[\begin{array}{c} \textit{a} \\ \{\{\text{sbj 3 sg m|wa}\}\} \end{array}\right]\!\!\right]\right)$:	⟨*a-sivyovi*taka, σ⟩

5 Summary and Discussion

The framework exemplified here distinguishes three kinds of exponence: intrinsic, positional, and conflated. It embodies two novel assumptions about the definition of a language's inflectional morphology.

First, *rules of inflectional exponence are resolved into two parts—an exponence declaration and a sequencing rule*; the former represents the intrinsic content expressed by an affix, and the latter accounts for its position in a word form's morphology together with any special content associated with its appearance in that position. The separation of exponence declarations from sequencing rules makes it possible for the same affix to be situated in more than one way (by more than one sequencing rule); this is a desirable consequence, since it is clear that an inflectional affix such as ⟦*vi*, {{ki|vi 3 pl}}⟧ may participate in a lexeme's morphology in more than one way, and may express additional, positional content depending on how it participates in it.

Second, *exponence declarations may be conflated to form more complex exponence declarations*. In general, both the form and the content of a conflation are deducible from those of its parts. But the conflation of two affixes has the effect of extending the range of morphotactic contexts in which one or both affixes may appear. For example, the sequencing rules for the noun-class concord ⟦*vi*, {{ki|vi 3 pl}}⟧ place this affix in the prefixal affix positions −1 and −3; but the sequencing rule for the conflation of ⟦*vi*, {{ki|vi 3 pl}}⟧ with ⟦*o*, {{rel}}⟧ places it suffixally in position ±2; and the sequencing rule for the conflation of ⟦*si*, {neg −tense}⟧ with ⟦*vyo*, {{rel ki|vi 3 pl}}⟧ places it prefixally in position ±2. Thus, the sequencing rules and the conflated exponence declarations in the proposed analysis allow ⟦*vi*, {{ki|vi 3 pl}}⟧ to appear (either alone or as part of a conflation) in four different morphotactic positions, expressing additional, positional content in two of these (object agreement in position −1 and subject agreement in position −3), and participating in the expression of two different kinds of conflated content in position ±2 (expressing relative agreement in suffixal position and tense or negation together with relative agreement in prefixal position).

The proposed approach to modeling Swahili verb inflection makes it possible to say that each Swahili noun-class concord has a constant bit of content (of which it is an intrinsic exponent) but may express additional content according to its position or to its conflation with other affixes. In particular, this approach correctly entails that by default, a subject-agreement affix has the same form as the corresponding object-agreement affix, since these two sorts of affixes are based on a common pool of intrinsic exponence declarations. Moreover, it correctly entails four facts about the Swahili relative affixes:

a. *The relative prefixes have the same form as the relative suffixes*, because they are based on a common pool of intrinsic and conflated exponence declarations.
b. *The relative suffixes are mutually exclusive with the* TN *prefixes*, because their sequencing rules belong to the same rule block, and are therefore in paradigmatic opposition.
c. *The relative suffixes are mutually exclusive with the relative prefixes*, because the relative affixes only appear as prefixes as a consequence of being conflated with a TN prefix, and these conflations are themselves TN prefixes (hence mutually exclusive with the relative suffixes; see again (b)).
d. *A relative prefix is always adjacent to a* TN *prefix*, because a relative affix only appears prefixally as a consequence of being conflated with a TN prefix.

The morphological framework proposed here sheds new light on familiar problems. An example is the cluster of properties exhibited by verbal clitics in many languages. In French, argument clitic pronouns exhibit a puzzling array of combinatorial properties. When any of the third-person accusative pronouns in (16) appears with any of the third-person dative pronouns in (17), the former precedes the latter. These pronouns ordinarily precede their verbal host, as in (18a, b, d); in affirmative imperative sentences, however, they follow it, as in (18c). Nevertheless, the precedence relation between the pronouns in (16) and those in (17) remains the same whether they are proclitic or enclitic, as these examples show.

(16) Third-person accusative clitic pronouns
  a. ⟦*le*, {{acc 3 sg masc −refl}}⟧
  b. ⟦*la*, {{acc 3 sg fem −refl}}⟧
  c. ⟦*les*, {{acc 3 pl −refl}}⟧

(17) Third-person dative clitic pronouns
  a. ⟦*lui*, {{dat 3 sg −refl}}⟧
  b. ⟦*leur*, {{dat 3 pl −refl}}⟧

(18) a. *Jean le lui donne.*
'Jean gives it to him.'

b. *Jean ne le lui donne pas.*
'Jean doesn't give it to him.'

c. *Donnez-le-lui!*
'Give it to him!'

d. *Ne le lui donnez pas!*
'Don't give it to him!'

Numerous, diverse analyses have been proposed for these facts, none of them fully satisfactory. The analytic framework proposed here, however, affords a simple explanation for these facts. At the core of this analysis is a distinction between the set of (simple and conflated) proclitic pronouns and the set of (simple and conflated) enclitic pronouns: I refer to pronouns belonging to both sets as **ambiclitic** pronouns; every third-person clitic pronoun is therefore ambiclitic. The rule in (19) defines the conflated exponence declarations in (20). Each of the ambiclitic pronouns in (16), (17), and (20) is placed by means of two sequencing rules, as defined in (21). Each of the sequencing rules defined by (21a) is a default, subject to override by the more specific rule defined by (21b), according to which enclitic pronouns are positional exponents of the affirmative imperative. In this analysis, all sequences of argument clitic pronouns are treated as conflated pronouns, and each of the conflated pronouns in (20) preserves the order of its parts whether it is sequenced proclitically by a rule defined by (21a) or enclitically by a rule defined by (21b).[6]

(19) The conflation of a third-person accusative clitic pronoun with a third-person dative clitic pronoun is an ambiclitic pronoun.

(20) Conflated ambiclitic pronouns
⟦*le-lui*, {{acc 3 sg masc –refl} {dat 3 sg –refl}}⟧
⟦*la-lui*, {{acc 3 sg fem –refl} {dat 3 sg –refl}}⟧
⟦*les-lui*, {{acc 3 pl –refl} {dat 3 sg –refl}}⟧
⟦*le-leur*, {{acc 3 sg masc –refl} {dat 3 pl –refl}}⟧
⟦*la-leur*, {{acc 3 sg fem –refl} {dat 3 pl –refl}}⟧
⟦*les-leur*, {{acc 3 pl –refl} {dat 3 pl-refl}}⟧

6 Luís 2004 and Luís & Spencer 2005 discuss clitic clusters in European Portuguese that likewise preserve the same relative ordering in proclitic and enclitic positions; it is reasonable to hypothesize that there, too, this fact reflects the order of clitic conflation.

(21) a. For every proclitic pronoun $[\![X, \tau]\!]$, the sequencing rule is Pref($[\![X, \tau]\!]$).
b. For every enclitic pronoun $[\![X, \tau]\!]$, the sequencing rule is Suff($[\![X, \{\text{affirmative imperative}\} \cup \tau]\!]$).

This analysis can be easily extended to cover non-third-person pronouns, which have five peculiarities:

A. They have the same form whether they accusative or dative: *Jean nous voit* 'Jean sees us,' *Jean nous donne un livre* 'Jean gives a book to us.'
B. The first- and second-person singular clitic pronouns have distinct proclitic and enclitic forms: *Jean me regarde* 'Jean looks at me', *Regardez-moi!* 'Look at me!'
C. As argument pronouns, they don't combine with each other: **Jean nous te présente* 'Jean introduces us to you/you to us.'
D. As dative pronouns, they combine with the third-person accusative pronouns in (16), but as accusative pronouns, they do not combine with the third-person dative pronouns in (17): *Jean nous la présente* 'Jean introduces her to us,' but **Jean nous lui présente* 'Jean introduces us to her.' (This is the so-called 'Person Case Constraint'; Haspelmath 2004, Rezac 2010.)
E. Although they precede the pronouns in (16) in proclitic position, they follow them in enclitic position:[7] *Ne nous le donnez pas!* 'Don't give it to us!' but *Donnez-le-nous!* 'Give it to us!'

In order to account for (A), we need only say that the intrinsic exponence declarations for the non-third-person pronouns are underspecified with respect to case, as in (22). In order to account for (B), the exponence declarations for first- and second-person singular clitics depend on whether they are proclitic (22b) or enclitic (22c). In order to account for (C), we assume that as argument pronouns, there is no conflation among the clitics in (22).

7 While peculiarity (E) is accurate for prescriptive French, spoken French sometimes deviates from (E): *Donne-moi-le!* For present purposes, I focus on the prescriptive patterns because they are the familiar subject of a large body of linguistic literature (e.g. Emonds 1975, 1978; Fiengo & Gitterman 1978; Kayne 1991; Miller & Sag 1997; Morin 1979a,b, 1981) and therefore provide a convenient basis for exemplifying the theoretical approach proposed here. That said, I see no reason to doubt that this theoretical approach will afford analyses of spoken French that are closely analogous in their simplicity to those that it affords for prescriptive French.

(22) a. Non-third-person ambiclitic pronouns
i. ⟦*nous*, {{1 pl}}⟧
ii. ⟦*vous*, {{2 pl}}⟧

b. Non-third-person proclitic pronouns
i. ⟦*me*, {{1 sg –nom}}⟧
ii. ⟦*te*, {{2 sg –nom}}⟧

c. Non-third-person enclitic pronouns
i. ⟦*moi*, {{1 sg –nom}}⟧
ii. ⟦*toi*, {{2 sg –nom}}⟧

In order to account for (D), we allow conflation between members of (22) and members of (16), but not between members of (22) and members of (17). And in order to account for (E), we assume the conflation rules in (23), which define the conflations in (24). The sequencing of clitic pronouns is then achieved by the sequencing rules given earlier in (21).

(23) a. The conflation of a non-third-person proclitic pronoun with a third-person accusative clitic pronoun is a proclitic pronoun.
b. The conflation of a third-person accusative clitic pronoun with a non-third-person enclitic pronoun is an enclitic pronoun.

(24) a. Conflated proclitic pronouns
⟦*me le*, {{acc 3 sg masc –refl} {dat 1 sg}}⟧
⟦*me la*, {{acc 3 sg fem –refl} {dat 1 sg}}⟧
⟦*me les*, {{acc 3 pl –refl} {dat 1 sg}}⟧
⟦*te le*, {{acc 3 sg masc –refl} {dat 2 sg}}⟧
⟦*te la*, {{acc 3 sg fem –refl} {dat 2 sg}}⟧
⟦*te les*, {{acc 3 pl –refl} {dat 2 sg}}⟧
⟦*nous le*, {{acc 3 sg masc –refl} {dat 1 pl}}⟧
⟦*nous la*, {{acc 3 sg fem –refl} {dat 1 pl}}⟧
⟦*nous les*, {{acc 3 pl –refl} {dat 1 pl}}⟧
⟦*vous le*, {{acc 3 sg masc –refl} {dat 2 pl}}⟧
⟦*vous la*, {{acc 3 sg fem –refl} {dat 2 pl}}⟧
⟦*vous les*, {{acc 3 pl –refl} {dat 2 pl}}⟧

b. Conflated enclitic pronouns
⟦*le-moi*, {{acc 3 sg masc –refl} {dat 1 sg}}⟧
⟦*la-moi*, {{acc 3 sg fem –refl} {dat 1 sg}}⟧
⟦*les-moi*, {{acc 3 pl –refl} {dat 1 sg}}⟧
⟦*le-nous*, {{acc 3 sg masc –refl} {dat 1 pl}}⟧
⟦*la-nous*, {{acc 3 sg fem –refl} {dat 1 pl}}⟧
⟦*les-nous*, {{acc 3 pl –refl} {dat 1 pl}}⟧

According to this analysis, a French verb form combines with at most a single argument clitic pronoun, which may be either simple (e.g. *me*) or conflated (*me le*). The intrinsic exponence declarations ((16), (17) and (22)) specify the content that a clitic expresses in all of its uses. The conflation rules ((19) and (23)) determine the relative order of argument clitic pronouns. The sequencing rules in (21) determine a (simple or conflated) clitic pronoun's order with respect to the verb stem to which it is affixed and specify the positional content ('affirmative imperative') of enclitic pronouns.

As this example shows, a morphological theory that distinguishes exponence declarations from sequencing rules and that allows the conflation of affixes affords new, more explanatory means of accounting for the phenomena of variable affix ordering and morphotactic conditioning.

References

Ashton, E.O. 1944. *Swahili grammar*. Essex: Longman.

Emonds, Joseph. 1975. A transformational analysis of French clitics without positive output constraints. *Linguistic Analysis* 1. 3–24.

Emonds, Joseph. 1978. The verbal complex V'–V in French. *Linguistic Inquiry* 9. 151–175.

Fiengo, Robert & Martin R. Gitterman. 1978. Remarks on French clitic order. *Linguistic Analysis* 4. 115–147.

Gazdar, Gerald, Ewan Klein, Geoffrey Pullum & Ivan Sag. 1985. *Generalized Phrase Structure Grammar*. Cambridge, MA: Harvard University Press.

Haspelmath, Martin. 2004. Explaining the Ditransitive Person-Role Constraint: A usage-based approach. *Constructions* 2, www.constructions-online.de. (29 April, 2012.)

Kayne, Richard S. 1991. Romance clitics, verb movement, and PRO. *Linguistic Inquiry* 22. 647–686.

Luís, Ana. 2004. *Clitics as morphology*. Colchester: University of Essex dissertation.

Luís, Ana & Andrew Spencer. 2005. A paradigm function account of 'mesoclisis' in European Portuguese. In Geert Booij & Jaap van Marle (eds.), *Yearbook of Morphology 2004*, 177–228. Dordrecht: Springer.

Miller, Philip & Ivan A. Sag. 1997. French clitic movement without clitics or movement. *Natural Language and Linguistic Theory* 15. 573–639.

Morin, Yves-Charles. 1979a. La morphophonologie des pronoms clitiques en français populaire. *Cahiers de Linguistique* 9. 1–36.

Morin, Yves-Charles. 1979b. More remarks on French clitic order. *Linguistic Analysis* 5, 293–312.

Morin, Yves-Charles. 1981. Some myths about pronominal clitics in French. *Linguistic Analysis* 8. 95–109.

Rezac, Milan. 2010. Ineffability through modularity: Gaps in French clitic clusters. In Matthew Baerman, Greville G. Corbett & Dunstan Brown (eds.), *Defective paradigms: Missing forms and what they tell us* [*Proceedings of the British Academy* 163], 151–180. Oxford: Oxford University Press.

Stump, Gregory T. 1993. Position classes and morphological theory. In Geert Booij & Jaap van Marle (eds.), *Yearbook of Morphology 1992*, 129–180. Dordrecht: Kluwer.

Stump, Gregory T. 2001. *Inflectional morphology*. Cambridge: Cambridge University Press.

CHAPTER 2

Verb Morphology and Conjugation Classes in Dunan (Yonaguni)

Thomas Pellard and Masahiro Yamada

1 Dunan and the Other Japonic Languages

1.1 *Japanese and the Other Japonic Languages*

Most Japonic languages have a relatively simple and transparent morphology.[1] Their verb morphology is usually characterized by a highly agglutinative structure that exhibits little morphophonology, with only a few conjugation classes and a handful of irregular verbs. For instance, Modern Standard Japanese has only two regular conjugation classes: a consonant-stem class (C) and a vowel-stem class (V). These two classes are further cross-classified by the accented vs. unaccented prosodic criterion (Table 2.1), but most of the allomorphy can easily be accounted for by phonotactic constraints: when two consonants or two vowels would cluster, deletion occurs ($*C_iC_j \rightarrow C_i$, $*V_iV_j \rightarrow V_i$).

TABLE 2.1 *Verb conjugation classes in Modern Standard Japanese*

	C		V	
	unaccented 'work'	*accented* 'swim'	*unaccented* 'align'	*accented* 'get up'
PRESENT	*hatarak-u*	*oyóg-u*	*narabe-ru*	*okí-ru*
NEGATIVE	*hatarak-anai*	*oyog-ánai*	*narabe-nai*	*okí-nai*
CONDITIONAL	*hatarak-éba*	*oyóg-eba*	*narabe-réba*	*okí-reba*
HORTATIVE	*hatarak-óo*	*oyog-óo*	*narabe-yóo*	*oki-yóo*
IMPERATIVE	*hatarak-é*	*oyóg-e*	*narabe-ró*	*okí-ro*
MEDIAL	*hatarai-te*	*oyói-de*	*narabe-te*	*óki-te*

1 See for example the descriptions in Shibatani (1990), Shimoji & Pellard (2010), Heinrich et al. (2015).

 | DOI 10.1163/9789004342934_004

Japanese is close to being a canonical agglutinative language, which explains why virtually all studies on Japanese verb morphology have adopted one of the various frameworks of morpheme-based morphophonology, like structuralist morphophonemics (Bloch 1946, Martin 1952), classical generative phonology (McCawley 1968), or optimality theory (Ito & Mester 2004). Though studies on other Japonic languages often follow the same vein,[2] the case of the Dunan (Yonaguni) language cannot easily be treated in the same way.

1.2 *Dunan (Yonaguni)*

Dunan (ISO 639–3 yoi), also called Yonaguni,[3] is a highly endangered Japonic language of the Southern Ryukyuan branch, spoken by approximately 400 speakers located on Yonaguni Island (Okinawa prefecture, Japan). In sharp contrast with its relatives, Dunan exhibits a unusually complex verb morphology for a Japonic language, mostly due to some drastic historical changes. The verb morphology of Dunan seems be the most complex one within the whole Japonic family, and a systematic description challenges in several interesting ways morpheme-based approaches.

The present work is based on original data gathered by the authors during fieldwork on Yonaguni Island between 2010 and 2014. The analysis relies on a morphological database containing more than 500 verbs obtained through either elicitation or analysis of oral text recordings. In order to test our understanding and analysis of Dunan verb morphology, we fully formalized our description and implemented it as a finite state transducer with the XFST software (Beesley & Karttunen 2003). The transducer obtained can be used for both generation and recognition. As a generator, it generates the complete paradigm of any verb, i.e. the set of all its possible inflectional forms together with their morphosyntactic labels. When used as a morphological analyzer, the transducer can output all morphosyntactic possible analyses of an unlabeled inflected form, i.e. its morphosyntactic features and the identity and conjugation-class membership of the underlying verbal lexeme. Preliminary tests

2 E.g. Ashworth (1973) on Shuri Okinawan, Lawrence (1990) on Nakijin Okinawan, or Arimoto (2001) on Dunan.

3 See Yamada et al. (2013) and Yamada et al. (2015) for a more detailed grammatical overview. The practical orthography adopted in this study uses the following conventions: *Ch* = [Cʰ]; *CC* = [Cˀ]; *y* = [j]; *r* = [ɾ]; *c* = [ts]; *c* → [tɕ], *s* → [ɕ], *h* → [ç] / __ {*i*,*y*}; *h* → [ɸ] ~ [ʍ] / __ {*u*,*w*}; *n* = homorganic / __ C, [ŋ] / __ #; X = any string including zero. Tone categories are indicated by an accent on the first vowel of a word (´ = High tone, ` = Low tone, ^ = Falling tone).

run on a small corpus of transcribed texts and elicited data revealed no error in our analysis.

The following presents an outline of the verb morphology of Dunan, limited to the basic synthetic forms of regular verbs. Focus is put on the partition of verbs into classes and its morphomic pattern. Three morphomic factors are identified as partitioning verbs into conjugation classes: stem alternation, suffix allomorphy, and metatony. The resulting system of paradigm classes is found to be opaque and to show little interpredictability between paradigm cells, i.e. few reliable inferences can be made from one inflected form about other forms. Morpheme-based approaches are not well-suited for the analysis of Dunan's verb morphology, which rather calls for an abstractive *Word and Paradigm* approach.

2 Verb Morphology in Dunan: Overview

2.1 *Verb Template*

The full morphological template of Dunan synthetic verb forms is illustrated in Table 2.2. We will focus in the following on the basic forms of regular verbs only and on the combination between a root and the first suffix since it is the main locus of complexity.

TABLE 2.2 *Verb template of Dunan*

0	1	2	3	4	5
			∅	PRESENT PAST	INDICATIVE PARTICIPLE
Root	CAUSATIVE	PASSIVE	NEGATIVE PERFECT	PAST ∅	CIRCUMSTANTIAL CONDITIONAL
	∅	∅	∅	∅	IMPERATIVE PROHIBITIVE HORTATIVE MEDIAL

2.2 *Conjugation Classes*

In Dunan, not all verbs exhibit the same morphological patterning in inflection, and verbs can thus be partitioned into several conjugation classes. These classes are not predictable from the phonological properties of verb roots, though there is indeed a correlation between conjugation classes and the segmental shape of roots. That correlation has been largely obscured by drastic historical changes, like the merger of high and mid vowels (**e* > *i*, **o* > *u*), vowel syncope in several environments, palatalization and subsequent deaffrication of velars before **i* (e.g. **k* > *t* / _*i*), or the univerbation of stems with a stative auxiliary. Compare for instance the alternations within the partial paradigms of the verbs 'blow' and 'fish' (< *'feed') in Dunan with the relative simplicity of their Modern Japanese cognates presented in Table 2.3.

As will be argued below, the conjugation classes of Dunan are not predictable from syntactico-semantic properties either. Identifying the class membership of a verb and its conjugation pattern is thus not straightforward and requires knowledge of several inflected forms.

The exact number of classes required in order to account for all patterns depends on the number and the degree of abstractness of the morphophonological rules one is ready to allow. In any case, conjugation classes are the product of the interaction of three independent factors: stem allomorphy, suffix allomorphy, and tone alternation (metatony).

3 Verb Stems

Whereas canonical inflectional classes are based on different sets of affixes (Corbett 2009), in Dunan, classes are primarily based on patterns of stem

TABLE 2.3 *Comparison between Dunan and Japanese verbs*

	'blow'		'fish', 'feed'	
	Dunan	Japanese	Dunan	Japanese
PRESENT	*kkùn*	*fúku*	*ppàn*	*kuwásu*
PAST	*ttìtan*	*fúita*	*ppàtan*	*kuwáshita*
IMPERATIVE	*kkì*	*fúke*	*ppài*	*kuwáse*
MEDIAL	*ttî*	*fúki*	*ppâsi*	*kuwáshi*

alternation rather than suffix allomorphy. Such alternations cannot be reduced to an epiphenomenon due to phonological readjustment rules, but they must be specified within the morphology. The exact number and analysis of these alternations depends on the segmentation procedure adopted, but the following account tries to maximize consistency and to keep balance between stem allomorphy and suffix allomorphy.

3.1 *Phonology of Stem Alternations*

Any regular verb exhibits up to three segmentally distinct stem forms in Dunan.[4] Such stems often exhibit a reduction/augmentation pattern, and it is generally difficult if not impossible to posit a single underlying base (Table 2.4).

For instance, velar-stem verbs like *sag·* 'tear' in Table 2.4 exhibit an alternation between a velar and a dental consonant: X*k·* ⟷ X*t·*, X*g·* ⟷ X*t·*, X*ŋ·* ⟷ X*d·* (Table 2.5). This alternation cannot be accounted for by a simple phonological rule of the type [DORSAL] → [CORONAL] / _ [CORONAL], where the coronal context stands for either *i* or *y*. Even though a dental stem appears in front of the medial verb suffix *-i*, e.g. *sat-i*, it fails to appear before the imperative (*-i*) and conditional (*-ya*) suffixes, e.g. *sag-i*, *sag-ya* (Table 2.5).

TABLE 2.4 *Phonology of stem alternations*

'see'	'tear'	'raise'	'rest'	'teach'	'dry'	'become'	'make'	'slip'
nn·	*sag·*	*kkan·*	*dug·*	*nar·*	*hw·*	*n·*	*kkw·*	*ngur·*
	sat·	*kkana·*	*dugw·*	*nara·*	*hu·*	*na·*	*kku·*	*nguri·*
			dugu·	*naras·*	*hus·*	*nar·*	*kkur·*	*ngurir·*

4 See section 5 for suprasegmental alternations. Here stems are followed by a raised dot (·) to indicate they are bound forms requiring a suffix, and not a hyphen (-) in order to clearly distinguish them from affixes.

TABLE 2.5 *Alternation between velars and dentals in verb stems*

	k X = *sun* 'pull'	*g* X = *sa* 'bloom'	*ŋ* X = *ha* 'deal'
CONDITIONAL	X*kya*	X*gya*	X*ŋya*
IMPERATIVE	X*ki*	X*gi*	X*ŋi*
PAST	X*titan*	X*titan*	X*ditan*
MEDIAL	X*ti*	X*ti*	X*di*
PERFECT	X*tyan*	X*tun*	X*dyan*

TABLE 2.6 *Morphologically conditioned alternations*

	X = *kh* 'do'	X = *ngara* 'wet'
IMPERATIVE	X*iri*	X*i*
MEDIAL	X*i*	X*si*
CONDITIONAL	X*irya*	—
PERFECT	X*yan*	X*syan*

The alternation between velars and dentals is thus not phonologically motivated and is not triggered by an automatic mechanism ensuring well-formedness conditions. It is the result of a diachronic phonological change which has been morphologized, i.e. it results from the palatalization and deaffrication of velars before **i* and the following raising of mid vowels (**e* > *i*).

Morphologically conditioned alternations are also found elsewhere (Table 2.6), like for example in sigmatic verbs,[5] whose imperative is of the form X-*i*, in contrast with their X*s-i* medial form. Both the imperative and the medial are marked by a suffix -*i*, but a different stem is used in each case. Similarly, for *i*-rhotic[6] verbs, the imperative (X*ir-i*) differs from the medial (X-*i*), and the conditional (X*ir-ya*) from the perfect (X-*ya*-), but they are differentiated by the shape of the stem rather than by that of the suffix.

5 Those verbs that exhibit a *s* ~ ∅ alternation.

6 Those verbs that exhibit a *r* ~ ∅ alternation and have a vowel *i* before the *r*.

3.2 *Patterns of Stem Alternation*

As stated above, the number of distinct stem forms per verb does not exceed three. However, the distribution of stems within paradigms is not uniform for all verbs (Table 2.7). For instance, in the case of verbs with three stems, the shortest stem usually appears only in the perfect cell, but not in the case of sigmatic verbs, where the same stem is shared by the hortative, medial, and perfect cells. On the other hand, for sigmatic verbs, the shortest stem is used in the negative cell only, while for most other classes the negative shares the same stem as the present and imperative forms.

The number of stems needed to account for the whole pattern of stem alternations across the different classes amounts to seven, i.e. more than twice the number of distinct stem forms for any verb (Table 2.7). This leads to the unexpected conclusion that there are almost as many stems as basic forms.

3.3 *Status of Verb Stems*

As illustrated above, the pattern of stem alternations of a verb depends on its conjugation class. Verbs from different classes can thus share one phonologically similar stem, while having completely different other stems (Table 2.4, 2.7). The distribution of stems within a paradigm is also class-specific. Obviously, it does not depend on phonological properties, but it is not

TABLE 2.7 *Patterns of stem alternation and their relative type frequency*

		X = *that* 'stand'	X = *sun* 'pull'	X = *ng* 'get wet'	X = *hud* 'grow'	X = *kk* 'make'	X = *kh* 'do'	X = *h* 'eat'	X = *dug* 'rest'	X = *mag* 'cook'	X = *ut* 'drop'	X = *h* 'dry'
1	NEG	X	X*k*	X*ar*	X*ur*	X*ur*	X*ir*	X	X	X	X	X*w*
2	PRES	X	X*k*	X*ar*	X*ur*	X*ur*	X*ir*	X	X	X*a*	X*u*	X*u*
2	COND	X	X*k*	X*ar*	X*ur*	X*ur*	X*ir*	–	–	–	–	–
3	IMP	X	X*k*	X*ar*	X*ur*	X*ur*	X*ir*	X*a*	X*u*	X*a*	X*u*	X*u*
3	CIRC	X	X*k*	X*ar*	X*ur*	X*ur*	X*ir*	X*a*	X*u*	X*a*	X*u*	X*u*
4	PROH	X	X*k*	X*a*	X*u*	X*u*	X*i*	X	X	X*a*	X*u*	X*u*
5	PAST	X	X*t*	X*a*	X*u*	X*u*	X*i*	X*a*	X*u*	X*a*	X*u*	X*u*
6	HORT	X	X*t*	X*a*	X*u*	X*u*	X*i*	X*a*	X*u*	X*as*	X*us*	X*us*
6	MED	X	X*t*	X*a*	X*u*	X*u*	X*i*	X*a*	X*u*	X*as*	X*us*	X*us*
7	PERF	X	X*t*	X	X	X*w*	X	X	X*w*	X*as*	X*us*	X*us*
	Freq.	10.60%	12.4%	14.44%	1.13%	5.64%	33.62%	2.71%	1.58%	14.67%	2.71%	0.45%

regulated by morphosyntactic features either. Depending on the verb class, stems will partition the paradigm cells in different sets, like {negative, present, prohibitive}, {imperative, circumstantial, past, hortative, medial}, {perfect} for vocalic verbs,[7] but {negative}, {present, imperative, circumstantial, prohibitive, past}, {hortative, medial, perfect} for sigmatic verbs. Such partitions are unmotivated since forms in the same set do not necessarily share common morphosyntactic features. It is even true for the imperative and the circumstantial, for which the same stem is used for all verbs while the two have little in common from a grammatical point of view.

The distribution of stems is purely morphological, and stems themselves underlie sets of forms that do not realize any coherent set of morphosyntactic features. The unavoidable conclusion is that in Dunan, stems are meaningless and are thus purely morphological objects, i.e. they are *morphomic* (Aronoff 1994).

4 Suffix Allomorphy

4.1 *Overview*

Suffix allomorphy is much less developed than stem allomorphy in Dunan verb morphology. A few cases of truncation of identical vowel sequences (V_i-$V_i \rightarrow V_i$) are found, like in the case of the medial form of *i*-rhotic verbs, where only one *i* surfaces from the concatenation of an *i*-ending stem with the suffix *-i*, but these are rather trivial.

Allomorphy is found in the perfect ({*ya*, *a*, *yu*, *u*}), circumstantial ({*uba*, *iba*, *ba*}), prohibitive ({*unna*, *nna*}), past ({*ita*, *ta*}), and present ({*u*, ∅}) suffixes. The choice of the allomorph can be shown to be independent of both phonological and morphosyntactic properties and to depend on purely morphological features.

4.2 *Conditioning*

Like stem alternation, suffix allomorphy is not phonologically conditioned in Dunan, despite what things look like at first sight. For instance, the choice between the two allomorphs *-uba* and *-iba* of the circumstantial suffix seems at first sight to be amenable to a phonological treatment, since they are in complementary distribution. The allomorph *-uba* is found after consonant-ending stems only, while *-iba* appears after vowel-ending stems only. This would however overlook the important fact that the choice of the stem appearing before the suffix is idiosyncratic and cannot be accounted for by the phonology.

7 Those verbs that exhibit a vowel ~ ∅ alternation.

For *u*-rhotic verbs, a consonant-ending stem is used (X*ur*-*uba*), even though a vowel stem is available ({X*u*·, X*ur*·, X*w*·}), while for *u*-sigmatic verbs a vowel-ending stem (X*u*-*iba*) is found, and not one of their consonant-ending stem ({X*u*·, X*us*·, X_C·}). Moreover, there is yet another allomorph -*ba* used in combination with the past suffix (X-*ta*-*ba*). The same is true for other cases of suffix allomorphy, and there is thus an arbitrary co-indexing of stems and suffixes.

The case of the perfect suffix is even more telling, since otherwise homophonous verbs take different suffix allomorphs. For example, verbs such as *nirun* 'boil' and *nirun* 'get boiled', or *sagun* 'tear' and *sagun* 'bloom', have identical paradigms except for their perfect form, respectively *nyan* vs. *nyun*, and *satyan* vs. *satun*.

Suffix allomorphy does not depend on syntactico-semantic properties either. For the perfect suffix, the correlation with transitivity or agentivity observed is at best a tendency and has many counterexamples (1).

(1)	a.	'align'	*narabirun* ↔ *narabyan*	≠	'enter'	*irirun* ↔ *irun*
	b.	'forget'	*bacirun* ↔ *bacun*	≠	'receive'	*ugirun* ↔ *ugyan*
	c.	'fall'	*utirun* ↔ *utun*	≠	'overflow'	*andirun* ↔ *andyan*
	d.	'copulate'	*ccungun* ↔ *ccungyan*	≠	'enter'	*irirun* ↔ *irun*
	e.	'get dry'	*ccirun* ↔ *ccun*	≠	'get boiled'	*nirun* ↔ *nyun*

4.3 *Status of Suffix Allomorphy*

Suffix allomorphy is rather limited in Dunan, and it is thus usually not a good predictor of class membership. The choice of an allomorph is not always predictable from the shape of the stem, and which stem is used with which suffix is anyway idiosyncratic and depends on the conjugation class of a verb. Here again, the analysis needs to recourse to arbitrary co-indexing of stems and suffixes, and since the choice of suffix does not depend on phonological or syntactico-semantic features, it is morphomic.

5 Metatony

5.1 *Overview of the Tone System*

Dunan has a lexical tone system where melodies apply to whole word forms, including any following clitic. Dunan has a three-way opposition between a High (H), Low (L), and Falling (F) word tone.[8]

8 Phonetically, High tone words have a high pitch throughout except for the first syllable of polysyllabic words. Falling tone words are similar to High tone ones, except that they have a falling pitch on the last syllable if it is heavy and they can trigger downstep on a following word. Low tone words have a low pitch throughout.

5.2 *Analysis of Metatony*

Though tones in Dunan are lexically determined, some verbs exhibit tonal alternations within their paradigm (Uwano 2011). Six patterns of paradigmatic metatony can be distinguished, but these can be reduced to four: a single non-alternating class ({H, L, F}), and three L ⟷ F alternating classes (Table 2.8).

Since tones are melodies which apply to whole word forms, they cannot be segmented into a stem part and an affixal part. Moreover, metatonic forms do not always share a common segmental stem form (Table 2.8, 2.9), like the negative (e.g. *ûm-anun* F) vs. past (e.g. *ûmu-tan* F) vs. perfect (e.g. *ûmw-an* F).

Putting the burden of metatony on stems would only lead to increase the number of stems: a new 'hortative' stem would be required, and as a result, only a few forms would share a common stem. Table 2.10 illustrates the increase in the number of stems resulting from attributing metatony to stems. For example, ascribing the L ↔ F alternation to stems leads to double the number of stems posited for a verb like 'tear', whose original stem set {*sag·*, *sat·*} would become {*sàg·*, *sâg·*, *sàt·*, *sât·*}.

TABLE 2.8 *Patterns of metatony and their type frequency*

	I *hún* 'eat'	II *nàgarirun* 'flow'	III *ndàn* 'get out'	IV *nnùn* 'watch'	V *ùmun* 'think'	VI *ân* 'be'
IMP						
PROH					L	
HORT				L		
CIRC			L			
PRES	H	L				F
PAST						
NEG					F	
PERF			F	F		
MED						
Freq.	44.09%	14.98%	27.09%	13.25%	0.29%	0.29%

TABLE 2.9 *Superposition of stem alternation patterns and metatony*

		X = *that* 'stand'	X = *sun* 'pull'	X = *ng* 'get wet'	X = *hud* 'grow'	X = *kk* 'make'	X = *kh* 'do'	X = *h* 'eat'	X = *dug* 'rest'	X = *mag* 'cook'	X = *ut* 'drop'	X = *h* 'dry'
1	NEG	X	X*k*	X*ar*	X*ur*	X*ur*	X*ir*	X	X	X	X	X*w*
2	PRES	X	X*k*	X*ar*	X*ur*	X*ur*	X*ir*	X	X	X*a*	X*u*	X*u*
	COND	X	X*k*	X*ar*	X*ur*	X*ur*	X*ir*	–	–	–	–	–
3	IMP	X	X*k*	X*ar*	X*ur*	X*ur*	X*ir*	X*a*	X*u*	X*a*	X*u*	X*u*
	CIRC	X	X*k*	X*ar*	X*ur*	X*ur*	X*ir*	X*a*	X*u*	X*a*	X*u*	X*u*
4	PROH	X	X*k*	X*a*	X*u*	X*u*	X*i*	X	X	X*a*	X*u*	X*u*
5	PAST	X	X*t*	X*a*	X*u*	X*u*	X*i*	X*a*	X*u*	X*a*	X*u*	X*u*
6	HORT	X	X*t*	X*a*	X*u*	X*u*	X*i*	X*a*	X*u*	X*as*	X*us*	X*us*
	MED.	X	X*t*	X*a*	X*u*	X*u*	X*i*	X*a*	X*u*	X*as*	X*us*	X*us*
7	PERF	X	X*t*	X	X	X*w*	X	X	X*w*	X*as*	X*us*	X*us*

TABLE 2.10 *Metatony and stems*

'see'	'tear'	'raise'	'rest'	'teach'	'dry'	'become'	'make'	'slip'
nn· L	*sag*· L	*kkan*· L	*dug*· L	*nar*· F	*hw*· L	*n*· F	*kkw*· F	*ngur*· F
nn· F	*sag*· F	*kkan*· F	*dug*· F	*nara*· L	*hu*· L	*na*· L	*kku*· L	*nguri*· L
	sat· L	*kkana*· L	*dugu*· L	*naras*· L	*hus*· L	*na*· F	*kku*· F	*nguri*· F
	sat· F	*kkana*· F	*dugu*· F	*naras*· F	*hus*· F	*nar*· L	*kkur*· L	*ngurir*· L
			dugw· F			*nar*· F	*kkur*· F	*ngurir*· F
+1	+2	+2	+2	+1	+1	+2	+2	+2

5.3 *Status of Metatony*

Like in the case of segmental allomorphy, metatonic forms do not share a common set of syntactic or semantic features. There is indeed no synchronic motivation behind the fact that the perfect and medial forms always undergo metatony in verbs exhibiting alternation, and the generalization that if a verb's past form undergoes metatony then its negative form does too has no motivation (Table 2.8). Metatony cannot therefore be considered to be the exponent of any morphosyntactic feature but has to be analyzed as a morphomic phenomenon.

6 Opacity

6.1 *Conjugation Classes*

Some conjugation classes in Dunan are barely distinguished from others but differ by a single form only. That form is however not the same one across all classes. This is the case of the negative form, which differs between verbs with otherwise identical paradigms (Table 2.11). For some other classes, it is the perfect that is the only different form (Table 2.12), but the perfect is itself ambiguous regarding class membership, and some verbs crucially lack a perfect form. This makes Dunan's system of conjugation classes an opaque and diffuse system in the sense of Stump & Finkel (2013).

TABLE 2.11 *Opacity in negative forms*

	'drop'		'push down'
PRESENT	*utun*	≃	*thun*
IMPERATIVE	*utui*	≃	*thui*
CIRCUMSTANTIAL	*utuiba*	≃	*thuiba*
PROHIBITIVE	*utunna*	≃	*thunna*
PAST	*ututan*	≃	*thutan*
MEDIAL	*utusi*	≃	*thusi*
PERFECT	*utusyan*	≃	*thusyan*
NEGATIVE	*utanun*	≠	*thwanun*

TABLE 2.12 *Opacity in perfect forms*

	'boil' (v.t.)		'boil' (v.i.)	'tear'		'bloom'
CONDITIONAL	*nirya*	=	*nirya*	*sagya*	=	*sagya*
NEGATIVE	*niranun*	=	*niranun*	*saganun*	=	*saganun*
PRESENT	*nirun*	=	*nirun*	*sagun*	=	*sagun*
CIRCUMSTANTIAL	*niruba*	=	*niruba*	*saguba*	=	*saguba*
PAST	*nitan*	=	*nitan*	*satitan*	=	*satitan*
MEDIAL	*nii*	=	*nii*	*sati*	=	*sati*
PERFECT	*nyan*	≠	*nyun*	*satyan*	≠	*satun*

6.2 *Individual Forms*

Diagnostic forms determining conjugation class membership are themselves ambiguous when taken in isolation (Figure 2.1). The consequence is that no single form can serve as an optimal principal part, and inferring a full paradigm requires the knowledge of several forms. Due to this opacity, no form can consistently predict another one across all conjugation classes (Table 2.13).[9] The process of inferring one form from another is thus always confronted with multiple possibilities, 3.5 on the average, but figures can vary greatly. Inferring the perfect from the present is the less reliable (8.04 possibilities), but ambiguity exists even in the best cases, e.g. inferring the present from the circumstantial, 1.13 possibilities on the average.

6.3 *No Blur Principle*

The case of the perfect in Dunan is problematic for the *No Blur Principle*, which states that "within any set of inflectional affixal realizations for the same paradigmatic cell, no more than one can fail to identify inflection class unambiguously" (Carstairs-McCarthy 1994: 742). Figure 2.2 illustrates that in some cases, all but one of the different possible realizations of the perfect suffix fail to unambiguously predict the present form, and thus the conjugation class of a verb.

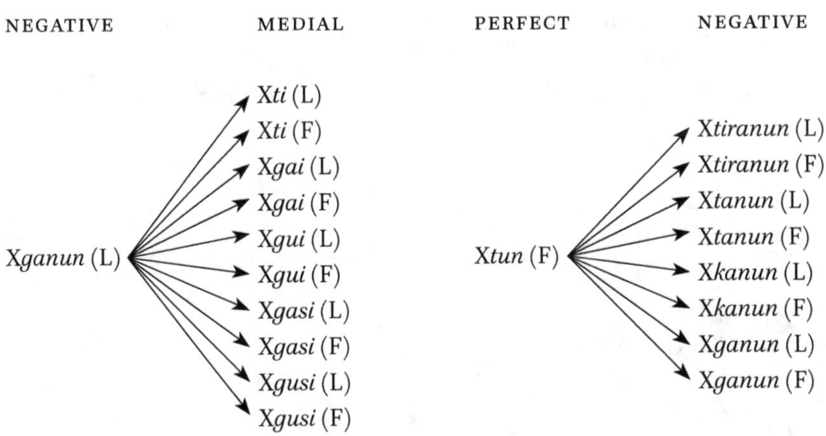

FIGURE 2.1 *Ambiguity of diagnostic forms.*

9 Calculations were done on a set of artificial verbs illustrating all possible combinations of stem alternations, suffix allomorphy and metatony. Table 2.13 thus represents the maximum opacity of the system, without taking into account neither systematic gaps nor frequencies.

TABLE 2.13 *Interpredictability of paradigm cells (e.g. for a given negative form, there are on the average 7.56 different possible corresponding perfect forms)*

	Predicted (mean number of forms)									
Predictor	CIRC	HORT	IMP	MED	NEG	PERF	PRES	PROH	PST	Mean
CIRC		1.33	1.20	2.40	2.16	3.60	1.13	1.13	1.80	1.84
HORT	3.13		3.13	1.80	6.00	5.52	2.73	2.47	3.48	3.53
IMP	1.73	1.87		3.36	2.04	4.56	1.13	1.67	3.12	2.44
MED	5.01	1.60	5.01		5.33	3.07	4.37	3.95	3.09	3.93
NEG	4.01	6.53	4.01	6.53		7.56	2.71	3.73	4.01	4.89
PERF	4.80	2.56	4.80	1.60	5.01		4.80	4.48	2.99	3.88
PRES	2.87	3.73	2.87	6.72	3.48	8.04		1.73	5.16	4.33
PROH	2.67	2.47	2.93	4.44	4.20	6.84	1.73		3.48	3.59
PAST	2.80	1.84	3.20	2.45	4.27	4.27	3.04	2.40		3.03
Mean	3.38	2.74	3.39	3.66	4.06	5.43	2.71	2.69	3.39	3.50

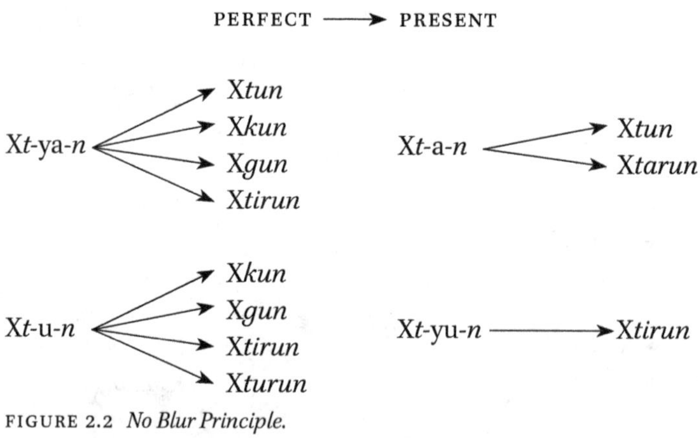

FIGURE 2.2 *No Blur Principle.*

6.4 *Systematic Homophony*

Some conjugation classes exhibit syncretism between some of their paradigm cells. This is for example the case for the simple past and perfect past, which are both realized as X*atan* for some classes, and also of the medial and imperative forms, both X*i* for some verbs.

Etymologically related transitivity pairs also exhibit reciprocal homophony between different cells of their paradigms, like in the case of *utun*, which can realize either the perfect of 'fall' (v.i) or the present of 'drop' (v.t.). Some perfect forms are ambiguously marked with a suffix *-u*, which with a different stem is an exponent of the present, like with 'bloom', present *sag-un* vs. perfect *sat-un*.

Such phenomena show that simply listing the presence of a stem or a word form for a verb is insufficient to identify that verb's conjugation class and infer its paradigm. The key information is to know what paradigm cell such a form realizes.

7 Which Model?

7.1 *Morpheme-based Morphophonology*

The data presented and analyzed above clearly shows that a morpheme-based morphophonological approach is not apt for describing the verb morphology of Dunan. Such an approach would only lead to posit rules or constraints that are unnatural and lack any generality because of their lexically or morphologically restricted character. Such unmotivated rules would be no more than diachronic changes disguised as synchronic rules.

Such an approach also fails on holistic *Gestalt* patterns, like 'tear' medial *sati*, imperative *sagi*, perfect *satyan*: *-i* cannot be an exponent of the medial category since it also marks the imperative, but the velar dental alternation cannot be an exponent of the medial either since it also characterizes the hortative, past, and perfect. It is the combination of the two, a specific stem with a specific suffix, that is meaningful.

A morpheme-based approach also generally fails to account for suppletive patterns. No phonological rule can relate 'be' *an* to its negative form *minun*. A unified treatment of the morphology, equally able to account for both regular and irregular forms, should be preferred.

7.2 *Root/Stem and Paradigm*

Several different alternative approaches to classical morpheme-based morphology have been proposed. For example the framework of stem spaces by Montermini & Bonami (2013) assumes that each lexeme comes equipped with a (possibly redundant) collection of stems indexed to paradigm cells. On the other hand, for Stump & Finkel (2013), word formation results from the addition of stem realization rules with exponence rules.

However, in the case of Dunan, the ten basic forms require at least eight different stems, and ten stems are needed in order to account for irregular verbs, so that the stem space is more or less equivalent with the basic paradigm. The stem realization rules would also be rather numerous. Plainly listing all the basic forms cannot thus be said to be less economical.

7.3 *Word and Paradigm*

An abstractive and associative model (Robins 1959, Matthews 1991, Blevins 2006) is particularly attractive in the case of Dunan. It assumes that forms do not exist in isolation but only as part of a structured system of interdependencies and implicational relations. It follows that roots, stems, and affixes are only *post hoc* abstractions over existing full forms, though they can have a utility as analytical or pedagogical devices. Such a model is thus based on exemplars and principal parts, and it relies on paradigmatic oppositions rather than syntagmatic relations.

In Dunan, there are usually fewer principal parts than stems for a given verb, especially if principal parts are conceived as *dynamic* sets (Stump & Finkel 2013), i.e as possibly differing according to the conjugation class. With principal parts, there is no more need to re-encode the information as combination rules and class diacritics, and we can capture implicative generalizations, either uni- or bidirectional ones, based on actual forms rather than stems:

(2) a. perfect X*yun* ⇒ circumstantial X*iruba*
b. medial X*ui* ⇔ perfect X*wan*

8 Conclusions

In the verb morphology of Dunan, neither stem alternations, suffix allomorphy, nor metatony can be viewed as clear exponents of syntactic or semantic features. They are thus better analyzed as being morphomic, i.e. purely morphological. Their cross-interactions define several different conjugation classes, which are also morphomic categories. Nevertheless, some secondary cues exist that reinforce the associative network of paradigmatic forms, like the limited correlation with phonotactics and semantics.

Stems and suffixes are not reliable indicators of class membership when taken isolatedly since they require "assembly instructions" (Blevins 2006): in order to know the shape of suffix, one needs to know the shape of the stem, but to know the shape of the stem, one needs to know the conjugation class, and to know the conjugation class, one needs to know the stems and suffixes of a verb. In some cases, like the perfect form, the choice of stem and suffix are even independent.

Dunan's verb morphology is characterized by the abundance of such holistic patterns and by the rather high opacity of its system. The most realistic hypothesis is that speakers memorize whole inflected forms (minimally, principal parts) as part of a network of interdependencies with an implicational structure. Unfortunately psycholinguistic tests are hardly feasible in the case of Dunan, a highly endangered language still spoken by elderly persons only. A more computational approach, based on entropy and complexity measures (Ackerman et al. 2009, Finkel & Stump 2009, Sagot & Walther 2011, Blevins 2013) is a more realistic goal for future research.

Appendix: Exemplary Paradigms

	X = *ndam* 'mince'	X = *thabu* 'fold'	X = *ba* 'spring'	X = *kk* 'use'	X = *pp* 'eat'	X = *nn* 'kill'	X = *n* 'fix'	X = *th* 'drop'
NEG	*Xanun*	*Xganun*	*Xganun*	*Xanun*	*Xanun*	*Xanun*	*Xanun*	*Xwanun*
PRES	*Xun*	*Xgun*	*Xgun*	*Xun*	*Xun*	*Xan*	*Xun*	*Xun*
COND	*Xya*	*Xgya*	*Xgya*	–	–	–	–	–
IMP	*Xi*	*Xgi*	*Xgi*	*Xai*	*Xui*	*Xai*	*Xui*	*Xui*
circ	*Xuba*	*Xguba*	*Xguba*	*Xaiba*	*Xuiba*	*Xaiba*	*Xuiba*	*Xuiba*
PROH	*Xunna*	*Xgunna*	*Xgunna*	*Xunna*	*Xunna*	*Xanna*	*Xunna*	*Xunna*
PAST	*Xitan*	*Xtitan*	*Xtitan*	*Xatan*	*Xutan*	*Xatan*	*Xutan*	*Xutan*
HORT	*Xindangi*	*Xtindangi*	*Xtindangi*	*Xaindangi*	*Xuindangi*	*Xasindangi*	*Xusindangi*	*Xusindangi*
MED	*Xi*	*Xti*	*Xti*	*Xai*	*Xui*	*Xasi*	*Xusi*	*Xusi*
PERF	*Xyan*	*Xtyan*	*Xtun*	*Xan*	*Xwan*	*Xasyan*	*Xusyan*	*Xusyan*

	X = *s* 'cut'	X = *nt* 'sit'	X = *hud* 'grow'	X = *cc* 'rub'	X = *h* 'go'	X = *nn* 'die'
NEG	*Xaranun*	*Xuranun*	*Xuranun*	*Xiranun*	*Xiranun*	*Xiranun*
PRES	*Xarun*	*Xurun*	*Xurun*	*Xirun*	*Xirun*	*Xirun*
COND	*Xarya*	*Xurya*	*Xurya*	*Xirya*	*Xirya*	*Xirya*
IMP	*Xari*	*Xuri*	*Xuri*	*Xiri*	*Xiri*	*Xiri*
CIRC	*Xaruba*	*Xuruba*	*Xuruba*	*Xiruba*	*Xiruba*	*Xiruba*
PROH	*Xanna*	*Xunna*	*Xunna*	*Xinna*	*Xinna*	*Xinna*
PAST	*Xatan*	*Xutan*	*Xutan*	*Xitan*	*Xitan*	*Xitan*
HORT	*Xaindangi*	*Xuindangi*	*Xuindangi*	*Xindangi*	*Xindangi*	*Xindangi*
MED	*Xai*	*Xui*	*Xui*	*Xi*	*Xi*	*Xi*
PERF	*Xan*	*Xwan*	*Xun*	*Xyan*	*Xyun*	*Xun*

References

Ackerman, Farrell, James P. Blevins & Robert Malouf. 2009. Parts and wholes: Implicative patterns in inflectional paradigms. In Blevins & Blevins (2009) 54–81.

Arimoto, Mitsuhiko. 2001. Ryūkyū Yonaguni Sonai hōgen no dōshi katsuyōkei no goi on'inron (1). *Yasuda Joshi Daigaku Kiyō* 29. 1–15.

Aronoff, Mark. 1994. *Morphology by itself: Stems and inflectional classes.* Cambridge: MIT Press.

Ashworth, David E. 1973. *A generative study of the inflectional morphophonemics of the Shuri dialect of Ryukyuan.* Ithaca: Cornell University dissertation.

Beesley, Kenneth & Lauri Karttunen. 2003. *Finite state morphology.* Stanford: CSLI Publications.

Blevins, James P. 2006. Word-based morphology. *Journal of Linguistics* 42(3). 531–573.

Blevins, James P. 2013. The information-theoretic turn. *Psihologija* 46(3). 355–375.

Blevins, James P. & Juliette Blevins (eds.). 2009. *Analogy in grammar: Form and acquisition.* Oxford: Oxford University Press.

Bloch, Bernard. 1946. Studies in Colloquial Japanese I: Inflection. *Journal of the American Oriental Society* 66(2). 97–109.

Carstairs-McCarthy, Andrew. 1994. Inflection classes, gender, and the principle of contrast. *Language* 70(4). 737–788.

Corbett, Greville G. 2009. Canonical inflectional classes. In Fabio Montermini, Gilles Boyé & Jesse Tseng (eds.), *Selected Proceedings of the 6th Décembrettes: Morphology in Bordeaux*, Somerville: Cascadilla Proceedings.

Finkel, Raphael & Gregory Stump. 2009. Principal parts and degrees of paradigmatic transparency. In Blevins & Blevins (2009) 13–53.

Heinrich, Patrick, Shinsho Miyara & Michinori Shimoji (eds.). 2015. *Handbook of the Ryukyuan languages: History, structure, and use.* Berlin; Boston: De Gruyter Mouton.

Ito, Junko & Armin Mester. 2004. Morphological contrast and merger: *Ranuki* in Japanese. *Journal of Japanese Linguistics* 20. 1–18.

Lawrence, Wayne P. 1990. *Nakijin phonology: Feet and extrametricality in a Japanese dialect.* Tsukuba: University of Tsukuba dissertation.

Martin, Samuel E. 1952. *Morphophonemics of Standard Colloquial Japanese.* Baltimore: Linguistic Society of America.

Matthews, Peter H. 1991. *Morphology.* Cambridge: Cambridge University Press.

McCawley, James D. 1968. *The phonological component of a grammar of Japanese.* The Hague; Paris: Mouton.

Montermini, Fabio & Olivier Bonami. 2013. Stem spaces and predictability in verbal inflection. *Lingue e linguaggio* XII(2). 171–190.

Robins, R.H. 1959. In defence of WP. *Transactions of the Philological Society* 58(1). 116–144.

Sagot, Benoît & Géraldine Walther. 2011. Non-canonical inflection: Data, formalisation and complexity measures. In Cerstin Mahlow & Michael Piotrowski (eds.), *Systems and frameworks in computational morphology: Second International Workshop, SFCM 2011*, 23–45. Berlin; Heidelberg: Springer.

Shibatani, Masayoshi. 1990. *The languages of Japan*. Cambridge: Cambridge University Press.

Shimoji, Michinori & Thomas Pellard (eds.). 2010. *An introduction to Ryukyuan languages*. Tokyo: ILCAA.

Stump, Gregory & Raphael. A. Finkel. 2013. *Morphological typology: From word to paradigm*. Cambridge: Cambridge University Press.

Uwano, Zendō. 2011. Yonaguni hōgen dōshi katsuyō no akusento shiryō (2). *Kokuritsu Kokugo Kenkyūjo Ronshū* 2. 135–164.

Yamada, Masahiro, Thomas Pellard & Michinori Shimoji. 2013. Dunan (Yonaguni)-go no kan'eki bunpō to shizen danwa shiryō. In Yukinori Takubo (ed.), *Ryūkyū rettō no gengo to bunka: Sono kiroku to keishō*, 291–324. Tokyo: Kuroshio shuppan.

Yamada, Masahiro, Thomas Pellard & Michinori Shimoji. 2015. Dunan grammar (Yonaguni Ryukyuan). In Heinrich et al. (2015) 449–478.

CHAPTER 3

Inflection Classes in Transitional Libyan Arabic (TLA)

Noura Ramli

1 Introduction

Traditionally, the morphology of Arabic has been treated as a non-concatenative type, having unique lexical elements including the consonantal and vocalic pattern which are also a feature of other language families. This approach has been adapted by what is widely known among most linguists as 'root and pattern' (McCarthy 1979, 1981; McCarthy & Prince 1990; Watson 2002). In one version of root and pattern morphology, exponents in morphological patterning are formalized by the prosodic units of foot, syllable, mora feet among others. In McCarthy (1979, 1981) the autosegmental phonology framework (Goldsmith 1976; cf. Harris 1941) was proposed to account for derivational and inflectional morphology of both verbal and nominal morphophonological alternations in Arabic along with other Semitic languages. This approach seemed to provide a plausible account for the morphological complexity of Arabic at that time. However, there are theoretical hurdles raised by this morpheme-based approach.

A radically different perspective on Semitic morphology known as the word-based/stem-based approach has been proposed. This approach promotes a uniform account of Arabic using a set of universal constraints. It is in many cases data driven, as is the case with this study. A small number of studies have argued that there is no need to invoke the separation of consonantal roots and vocalic elements when the morphological patterning in Arabic can be handled and effectively accounted for by the stem (McOmber 1995; Benmamoun 1999, 2003; Gafos 2003; Heath 1987, 1997, 2003; Ratcliffe 1998; Darden 1992; Camilleri 2014). For example, McOmber (1995) argues that the representation of the lexical entry in Arabic can be the CCVC stem and there is no need for the consonantal root representation. Similarly, Benmamoun (1999, 2003) is in favour of a stem-based model in which the imperfective verb form has underived status and serves as the verb default form and the base for lexeme formation processes. The stem account is also proposed for Arabic dialects such as Cairene

 | DOI 10.1163/9789004342934_005

(Darden 1992), Moroccan (Heath 1987). The inadequacy of the root and pattern approach is also demonstrated by Optimality Theory accounts of Semitic morphology (Ussishkin 1999, 2000; Gafos 2003).

The aim of this paper is to discuss the TLA verb complexity under a stem based-account in which the morphomic stems are hierarchically defined based on dependency relations, hence they are lexically defined classes. First, the general features and the verb forms of TLA are outlined in section 2. In section 3, I introduce the paradigm scheme for TLA verbs. The final section presents the TLA stem space and the dependency relations between stems of different verb series. Most of the studies have been concerned with the stem form and role of stems in word formation. However, none of the studies on Arabic has specifically addressed the issue of the morphomic status of the stem and the motivating factor(s) for stem allomorphy in Arabic verb inflectional paradigms. Furthermore, there is hardly any work investigating the role of stem allomorphy in classifying and determining implicative relations and ICs in the Arabic system.

2 TLA Verbal System

Previous analyses of Libya Arabic (LA) verb morphology using different frameworks provide a useful source of comparison for the account I present in section 3. Descriptions exist for Eastern Libyan Arabic (Owens 1984), Al-Jabal Libyan Arabic (Harrama 1993) and Tripoli Libyan Arabic (Elfitoury 1976). Owens (1984) identifies three main types of Libyan Arabic which include the Western dialect spoken mainly in the capital city, the Eastern dialect used in and around the region of Bengazi and finally the TLA spoken in Northwestern Libya in Misrata city and Sabah. TLA shares some features of the other dialects, but it also shows distinct features. As in many traditional descriptions of Semitic languages we can divide TLA verbs into two classes depending on whether they are formed from a triliteral (three-consonant) or a quadriliteral (four-consonant) root.[1] TLA verbs can also be categorised along a second dimension—the set of verb measures. The verb measures are traditionally defined as a systematically related set of derived verb forms, expressing distinctive grammatical features (causative reflexive, passive among others). Measure one (M1) is considered to be the underived verb form, and the CV

1 In this work, I will only consider the triliteral verbs that are relevant to the unique stem morphomic pattern found in TLA.

templates for the other measures are derived from this by consonantal root alternations, vocalic alternation and affixation (McCarthy 1979, 1981, 1986). I follow Hoberman and Aronoff (2003: 63) in assuming that each verb measure 'consists of an inner stem with fixed vocalism and a fixed prosody, surrounded by fixed affixes (if they are present), and in some cases different prosodies and vocalisms in different tenses, aspects, or moods. Consonantal roots are fitted to these complex patterns.' Each measure functions as 'an inflectional class, it dictates the phonological form of the verb' (Aronoff 1994: 127).[2] It is generally assumed that Libyan Arabic retains eight measures, M1 and non-M1 (M2, M3, M4, M5, M6, M7, M8) of the ten Standard Arabic (SA) verb measures for the triliteral verbs (Table 3.1).[3] In addition, TLA verbs can belong to one of four different phonologically defined verb series, the sound verbs, double verbs, hollow verbs and the defective verbs (Table 3.2).

TABLE 3.1 *TLA verb measure patterns*

Measure	example	template	Gloss
1	leʕab	CVCVC	'play'
2	ferraḥ	CVC_iC_iVC	'make sb. happy'
3	šaarik	CV_iV_iCVC	'share with sb.'
5	tekallim	$t(e)CV\ C_iC_iVC$	'talk with'
6	teʕaarik	$t(e)CV_iV_iCVC$	'argue'
7	inkesar	(i)nCVCVC	'be broken'
8	irtefaʕ	(i)CtVCVC	'be taken'
10	staʕmil	st(a)CCVC	'use'

2 Note that Aronoff is using the term 'inflectional class' in a very specific sense here, which is rather different from the sense adopted here, and in most of the literature.

3 Measure four (M4) (e.g. *ʔaxraj* 'expel') and measure nine (M9) (e.g. *ihmarra* 'become red') of SA are not attested in TLA.

TABLE 3.2 *The possible verb series for each measure*

Measure	Regular (strong)	Irregular (weak)		
	Sound	Double	Hollow	Defective
1	CVCVC ketab 'write'	CVC_iC_i ḥall 'open'	CVVC ġaab 'be absent'	CCVV jree 'run'
2	CVC_iC_iVC kallim 'talk'	$CVC_iC_iVC_i$ ḥabbib 'make sb. love'		CVC_iC_iV garra 'teach'
3	CVVCVC naašib 'harass'			CVVCV naada 'invite'
5	$tCVC_iC_iVC$ teʕallim 'learn'	$tCVC_iC_iVC_i$ teġaššiš 'feel upset'		$tCVC_iC_iV$ tewaffa 'die'
6	tCVVCVC teʕaarik 'argue'			tCVVCV telaaga 'meet'
7	inCVCVC inkeṣar 'be broken'	$inCVC_iC_i$ ingall 'be taken'	inCVVC inbaaʕ 'be sold'	inCCVV indʕee 'be infected'
8	iCtVCVC irtefaʕ 'be taken'	$iCtVC_iC_i$ irtadd 'be returned'	iCtVVC ixtaar 'choose'	iCtCVV intsee 'be forgotten'
10	stCVCVC staʕmil 'use'	$stCVC_iC_i$ steġall 'make use of'	stCVVC stefaad 'benefit'	stVCCV stanja 'appeal'

M1 can be classified as the simple (active or basic voice) verb form from which other verb patterns can be derived, but not necessarily all. For example, the verb *rabbij* 'sing traditional songs,' which belongs to M2, has no M1 counterpart **rebaj*. The main characteristics of M2 is doubling represented by a geminate second consonant and it can belong to any verb series except the hollow series. M3 tends to be limited to a small number of transitive verbs (Harrama 1993) and it can have the sound and defective series only. M5 can be derived from M2 by attaching a prefix (t-), conveying a passive or sometimes a reflexive meaning. M6 is based on M3 with the addition of a prefix (t-), signifying a reciprocal meaning. Both M7 and M8 can function as a medio-passive (inchoative) verbs. In fact, they have the same pattern as M1 with the addition of the prefix (n-) and the infix (-t-) respectively. The derivation of M10 involves prefixing (st-) to the root of the verb (usually a M1 verb) (ElFitoury 1976: 30–35). The sound series is characterised by the presence of three distinct consonants while the double series has geminate consonants word-medially or word-finally. The structure for the hollow series is based on a long stem vowel whereas the defective series ends in a vowel which alternates with a monophthongal ending *ee*[4] before consonant initial suffixes. It is also worth noting that in root and pattern based approaches, the 3SGM perfective form serves as the base form (McCarthy 1981; Guerssel & Lowenstamm 1996) whereas stem-based models consider the 3SGM imperfective CCVC to be the default form (Ratcliffe 1997; Benmamoun 2003; Heath 2003; Gafos 2003). I follow the stem-based works in regarding the imperfective as the default form. However, it is also possible to treat the 3SGM perfective form as the default form. In either case, the results seem to be the same in terms of stem dependency relations (cf. section 3).

Each verb measure and series can be inflected for different morphosyntactic properties (MPS). TLA verbs realize a number of MPS, some of them expressing grammatical meanings, others used in agreement processes. Verbs distinguish finite forms from non-finite forms (participles and masdars). The inventory of MPS for finite verb forms is summarized in Table 3.3:

4 In TLA defective verbs, the monophthong *ee* is possibly a diachronic process of *imala* (the raising of the stem final vowel *a* to *ee*). In JLA, the defective verbs have /a/ stem final vowel. By contrast, the *ee* in TLA double verbs is not a case of *imala*. In this work, I will refer to the two cases as a monophthong.

TABLE 3.3 *TLA MPS*

MP	values
ASPECT (ASP)	perf(ective), imperf(ective)[a]
TENSE (TNS)	fut(ure), non-fut(ure)
NUMBER (NUM)	SG, PL
PERSON (PER)	1, 2, 3
GENDER (GEN)	M, F

a A passive-type construction in TLA can be realized by both M7 and/or M8. Note, however, that the M7 paradigm of the verb *inbetah* 'lie down,' for example, can have active forms alongside passive forms. In other words, the M7 forms do not realize a voice opposition as such. For these reasons, we might want to say that there is, in fact, no inflectional voice opposition in TLA.

The future tense and imperative forms are only ever imperfective in aspect, and the forms of the future are derived by prefixation from the imperfective forms in a completely regular way, and therefore add nothing of interest to paradigm structure. Verbs agree with the subject in person (1st, 2nd, 3rd), number (SG/PL), and gender (M/F). In the imperative the verb only realizes 2nd person. The full set of MPs is realized by affixation and the same inflectional affixes for the same cells are used across different verb measures/series. TLA verb inflection involves suffixing and/or prefixing the verb stem. For instance, the inflection for the MPs in the perfective sub-paradigm is realized by suffixing the verb stem while verbs are inflected for imperfective by both prefixes and suffixes. In the imperative, agreement is realized by single suffix cumulating (2nd) person/gender. In M1 sound and defective series the imperfective and imperative forms also have a pre-formative vowel (FV). This is a vowel that precedes the verb stem and follows the imperfective prefix (e.g. *y-a-rkub* 'he rides', *y-i-kr-u* 'they hire'). TLA verbs have object pronoun suffixes in the perfective and imperfective paradigm.[5]

5 All object suffixes, except for the first person forms, can also be attached to nouns as possessive pronouns.

TLA is a fusional system, in which portmanteaux morphs are found in adjacent slots. Verbs agree with the subject NP and object NP in person number and gender. TLA morphotactics for linear order requires the negation affix/clitic *ma-* to precede the subject affix on the left edge of the stem and the perfective person number/gender subject suffixes precede object suffixes (Table 3.4):

TABLE 3.4 *The position classes of portmanteau morphs in TLA in* mayikitbulnaš *'They do not write to us' (following the formatting in Crysmann & Bonami 2012)*

1	2	3	4	5	6	7	8
Neg	Fut marker	Subj marker	**stem**	Subj marker	Ind obj marker	Obj marker	Neg
ma		y	(i) kitb	u	l	na	š

The 'templatic organization' in Table 3.4 is shared by all verb measures, thus, following Owens (1984), we provide the first segmentation account for TLA and suggest that all verbs inflect in the same manner, that is, there is unified exponence for all verbs (Table 3.5).

The affixes show a fusional pattern with regard to position class. The suffix /-t/ realizes the MPs of 1/2SGM in the perfective sub-paradigm. In the imperfective, the slot 3 prefix /t-/ realizes aspect and number. The two /t/ affixes are an example of homophony. Another observation about the TLA affix inventory is that the subject and object markers of 1PL are homophones, but occur in different slots: 5 and 7 respectively (Table 3.4).[6] The fact that affixes are unified across all the different verbs means that affixes are redundant and cannot be identifiers for IC membership. The next section will discuss the ICs in TLA based on the stem space framework.

6 The unified account of exponents might seem to suggest a simple segmentation procedure, but the FV and the monophthong can raise a 'segmentation problem.'

TABLE 3.5 *TLA affix inventory*

		Subject						Object[a]	
		Perfective		Imperfective		Imperative			
		SG	PL	SG	PL	SG	PL	SG	PL
1		-(i)t	-na[b]	n-	n-. . .-u			-ni	-na
2	M	-t	-tu	t[c]-	t-. . .-u	Ø	-u	-ik	-kum
	F	-ti	-tin	t-. . .-i	t-. . .-in	-i	-in	-ik	-kin
3	M	Ø	-u	y-	y-. . .-u			-a(a)(h)[d]	-hum
	F	-it ~ aat-	-in	t-	y-. . .-in			-ha	-hin[e]

a When object suffixes are added to a subject suffix the V or C of the first suffix undergoes lengthening (e.g. *kitib-u* 'they.M wrote,' cf. *kitib-uu-hin* 'they wrote them.F' and *kitib-in* 'they.F wrote' cf. *kitibinn-a* 'they wrote it.M'). According to Mitchell (1960: 386) the reason for this adjustment process is to minimize sequences of 'short open syllables.' Owens (1980) adds that the vowel change in the 3SGF suffix from high vowel /i/ to a low vowel /a/ when the 3SGM object, /-aat-a/, is added seems to suggest that this suffix vowel originates as /-at/ across the whole paradigm. In TLA, this suffix is realized with a long low vowel /-aat/ when followed by the object suffixes /a/ and /Vk/.

b The 1PL subject and 1PL object suffixes show affix homophony (Spencer 1991: 211–212).

c The prefix (t-) is subject to the phonological process of assimilation.

d Owens (1984: 93) points out that, with regard to the 3SGM.obj marker, 'in connected speech the lengthened vowel form without the *-h* is the more usual.'

e The 3PL.M.obj cannot combine with 2PL.F suffix **tin+hum* and instead the 2PLM subj and 3PLM.obj suffix combination (*tu-hum*) is used with feminine gender. However, the 2PLF and 3PLF suffixes (*tin-hin*) is a possible morph combination. This could be an example of an unmotivated affix syncretism, showing that the subject-object markers are part of an inflectional paradigm.

3 TLA Stem Space

The abstract paradigm schema consists of 24 cells as illustrated in Table 3.6. The schema represents all the permitted combinations of values. Each cell is generally filled by one unique inflected form, but in some cases we see syncretism, for instance, between the 1SG and the 2SGM cells in the perfective paradigm of all measures/series. This is illustrated by the following M1 forms: *ketab(i)t*

TABLE 3.6 *Paradigm schema for* TLA *verbs*

Mood	Indic		Imper
Asp	Perf	Imperf	
1SG			
1PL			
2SGM			
2SGF			
2PLM			
2PLF			
3SGM			
3SGF			
3PLM			
3PLF			

'I/you wrote' (sound series), *gult* 'I/you said' (hollow series), *maddeet* 'I/you passed' (double series), *nseet* 'I/you forgot' (defective series). Some irregular verbs have defective paradigms, that is, they systematically fail to fill certain cells. For example, the verb *y-ibi* 'he wants' lacks the perfective sub-paradigm, and this gap has to be filled periphrastically: *kan yi-bi* 'he had wanted.'

The aspectual contrast is sometimes accompanied by stem changes.[7] For example, in M1 sound verbs, the vowel pattern in the perfective 1st/2nd and 3SGM is *e-a* whereas the 3SGF/PL can have *i-i* or *u-u*.[8] In the imperfective counterparts, the FV is lexically determined. The identity of the stem vowels can be partially determined by phonological factors such as assimilation to the adjacent FV (*yu-rugd-u* 'they sleep,' but *ya-ṣugṭ-u* 'they fail' *ya-simʕ-u* 'they hear') or

7 In SA, by contrast, each measure can only have one vocalism within each sub-paradigm. For instance, *katab* 'he wrote' has the same vowel(s) for all the inflected forms in the perfective paradigm.

8 E.g. *ketab* ~ *kitib* 'write,' *resam* ~ *risim* 'draw,' *ʕegal* ~ *ʕigil* 'become mature/hide,' *gelaʕ* ~ *giliʕ* 'take off clothes/extract,' *ʕegad* ~ *ʕigid* 'tie,' *serag* ~ *sirig* 'steal,' *sekab* ~ *sikib* 'pour,' *seraḥ* ~ *siriḥ* 'wonder,' *feham* ~ *fihim* 'understand,' *felag* ~ *filig* 'burst,' *fesad* ~ *fisid* 'rotten,' *gesam* ~ *gisim* 'divide,' *geʕad* ~ *guʕud* 'stay,' *regad* ~ *rugud* 'sleep,' *zerag* ~ *zurug* 'go quickly for a mission,' *sekat* ~ *sukut* 'became silent,' *sekan* ~ *sukun 'live,' ṣebaḥ* ~ *ṣubuḥ* 'be in a certain state in the morning,' *feraḥ* ~ *furuḥ* 'be happy.'

being conditioned by the final stem consonant (ʕ, ḥ or *h*). Similarly, the other (non-sound) M1 verb series and some of the non-M1 verbs can also show intra-paradigmatic morphological changes in the form of stem ablaut and suffixation of the monophthongal ending *-ee*. An example of a full inflectional paradigm for each of the four M1 verb series and the M2 sound and defective series is presented in the appendix.

TLA verbal paradigms display a systematic stem variation that involves changes to the 'stem-shape' by altering the syllable structure, ablaut and addition of the monophthong *ee* extension (see below). These stem changes can be motivated by morphological factors (Maiden 2009, 2011; Bonami & Boyé 2002) and extramorphological factors including phonological (Anderson 2011), morphophonological (Spencer 2013; Maiden 2011) and morphosyntactic (Baerman & Corbett 2012). Following Boyé & Cabredo-Hofherr (2006), the stem change includes all patterns of stem-shape variations. The stem variation pattern will be represented by the metaparadigm in which every stem that marks a stem change has a different letter. It is worth noting that this representation is different from the 'stem-space' in which only indexed stems that are morphologically motivated are listed (Table 3.10). The reason for using this framework for classifying TLA stems is that it can determine the factors that control the change and provide a deeper understanding of the theoretical status of the stem.

To establish the characteristics of this framework, we consider the paradigm of the regular verb types in the language, the M1 sound series. For instance, the inflectional paradigm of the verb *ġesal* 'wash' shows the different stems across imperfective, imperative and perfective. The pattern of the change can also vary depending on whether the verb is marked by subject affixes or object suffixes as shown in the meta-paradigm of *ġesal* (Table 3.7).

The paradigms in Table 3.7 and Table 3.8 show that within the imperfective sub-paradigm with subject suffixes, the stems are in allomorphic relations and these same stems are used in the imperative.[9] Likewise, within the perfective sub-paradigm with subject markers, the 1/2 person forms share the same stem while the 3rd person forms show two stem variants. With regard to inflectional forms with object markers, stem change can be observed when an object marker is attached to any of the unsuffixed forms such as the 1SG, 2SGM, 3SG imperfective forms, and the 3SGM perfective form. However, no stem change is triggered when an object suffix is appended to the verb, but preceded by a subject marker. In addition, the listing of all possible stem changes is redundant

9 The imperative and the future forms are parasitic on the imperfective forms. Therefore, the same stems are selected in cells with the corresponding person, number, gender MPs.

TABLE 3.7 *The paradigm of M1 sound series* ġesal *'wash"*

MPS	Subject			3SGM.Obj	
	Imperfective	Imperative	Perfective	Imperfective	Perfective
1SG	n-uġsil		ġesal-t	y-uġsil-ni	ġesal-ni
1PL	n-uġusl-u		ġesal-na	y-uġsil-na	ġesal-na
2SGM	t-uġsil	uġsil	ġesal-t	y-uġusl-ik	ġusul-ik
2SGF	t-uġusl-i	uġusl-i	ġesal-ti	y-uġusl-ik	ġusul-ik
2PLM	t-uġusl-u	uġusl-u	ġesal-tu	y-uġsil-kum	ġesal-kum
2PLF	t-uġusl-in	uġusl-in	ġesal-tin	y-uġsil-kin	ġesal-kin
3SGM	y-uġsil		ġesal	y-uġusl-a	ġusul-a
3SGF	t-uġsil		ġusul-it	y-uġsil-ha	ġesal-ha
3PLM	y-uġusl-u		ġusul-u	y-uġsil-hum	ġesal-hum
3PLF	y-uġusl-in		ġusul-in	y-uġsil-hin	ġesal-hin

TABLE 3.8 *The stem metaparadigm of the verb* ġesal *'wash'*

MPS	Subject			3SGM.Obj	
	Imperfective	Imperative	Perfective	Imperfective	Perfective
1SG	A		C	A	C
1PL	B		C	A	C
2SGM	A	A	C	B	D
2SGF	B	B	C	B	D
2PLM	B	B	C	A	C
2PLF	B	B	C	A	C
3SGM	A		C	B	D
3SGF	A		D	A	C
3PLM	B		D	A	C
3PLF	B		D	A	C

TABLE 3.9 *The representative forms for* TLA *stem change*

Subject					3SGM.Obj	
Imperfective		Perfective			Perfective	
3SGM	3PLM	1PL	3SGM	3SGF	1PL	3SGM

due to the systematic predictability across paradigm cells (Montermini & Bonami 2011; Bonami & Boyé 2008; Boyé & Cabredo-Hofherr 2006). Therefore, to avoid the redundancy, we will conventionally choose the forms in Table 3.9 for each sub-paradigm as representatives of a set of cells with the same stem. The stem selected in these cells will be the basis for the pattern of stem change in each verb measure/series.[10]

For the imperfective sub-paradigm, the first representative cell is the 3SGM that represents every unsuffixed form and any form with a consonant-initial object suffix. The second selected cell is that of the 3PLM form which includes all forms with a vowel-initial subject or object suffix. Across the perfective sub-paradigm, the 1PL form shares the same stem with the 1SG and 2nd person forms. The reason for listing this form is that it is essential for illustrating the contrast between the stem patterns when a M1 double verb is marked with the 1PL.obj suffix. Although the 3SGF/PL forms share the same stem in all verbs, the 3SGM can have a variable pattern depending on the verb series and measure. Therefore, the cells of both 3SG forms will be represented. Finally, the representative paradigm will also list the stem changes of the unsuffixed perfective 3SGM form when the 3SGM and 1PL object markers are attached to the form. The reason for representing the 3SGM.obj sub-paradigm, is that it shows distinctive pattern of stem change distinguishing different sub-patterns within the same verb series. In TLA, the sound series of M1 is the most frequent verb form and it has multiple stems in the perfective and imperfective sub-paradigms while non-M1 verbs have less/no stem variation.

10 Note that the sub-paradigm represented by the 3SGM.obj form will not be represented with intransitive verbs, M7/8 verbs (e.g. *inkesar* 'it was broken') or M6 verbs (e.g. *telaaga* 'they met each other').

The stem change is morphologically induced and the verb inflectional complexity in TLA can be characterised by a stem space that consists of a set of stems—each of which is connected to a slot in the inflectional paradigm. The data in Table 3.10 show that all 1.2SG/PL perfective inflectional forms share the same stem in the slot whereas 3SG/PL perfective inflectional forms may select distinct stem slots: 3SGM slot and 3FSG/PL slot. The majority of the non-M1 verbs select the same stem for all the slots. I propose that the maximal stem space in TLA is derived from the maximum number of stem *ablaut changes* and monophthong changes displayed by the pattern of stem change, which

TABLE 3.10 *The stem change in the representative forms of the subj sub-paradigm across all measures*

verb	Gloss	M	Imperfective		Perfective		
			3SGM	3PLM	1PL	3SGM	3SGF
sound series							
getal	kill	M1	ugtil A	ugutl B	getal C	getal C	gutul D
ʕallim	teach	M2	ʕallim A	ʕallm A	ʕallim A	ʕallim A	ʕallm A
ʕaamil	deal	M3	ʕaamil A	ʕaaml A	ʕaamil A	ʕaamil A	ʕaaml A
teʕallim	learn	M5	teʕallim A	teʕallm A	teʕallim A	teʕallim A	teʕallm A
teʕaamil	deal	M6	teʕaamil A	teʕaaml A	teʕaamil A	teʕaamil A	teʕaaml A
insemaʕ	be heard	M7	insimaʕ A	insimiʕ B	insemaʕ C	insemaʕ C	insimiʕ B
irtefaʕ	be carried	M8	irtifaʕ A	irtifiʕ B	irtefaʕ C	irtefaʕ C	irtifiʕ B
staʕmil	use	M10	staʕmil A	staʕml A	staʕmil A	staʕmil A	staʕml A
defective series							
nsee	forget	M1	ansa A_i	ans A_j	nsee B_i	nsee B_j	ns B_k
bakka	upset	M2	bakki A	bakk A	bakkee A	bakka A	bakk A

verb	Gloss	M	Imperfective		Perfective		
			3SGM	3PLM	1PL	3SGM	3SGF
naada	invite	M3	naadi A	naad A	naadee A	naada A	naad A
temanna	wish	M5	temanna A	temann A	temannee A	temanna A	temann A
telaaga	meet	M6	telaaga A	telaag A	telaagee A	telaaga A	telaag A
inkiwee	cauterised	M7	inkiwi A	inkiw A	inkiwee A	inkiwee A	inkiw A
intisee	forgotten	M8	in(t)isa A	in(t)is A	in(t)isee A	in(t)isee A	in(t)is A
stanja	purify	M10	stanja A	stanj A	stanjee A	stanja A	stanj A
double series							
gall	take	M1	gill A	gill A	gallee B	gall B	gall B
ḥaddid	specify	M2	ḥaddid A	ḥaddid A	ḥaddid A	ḥaddid A	ḥaddid A
tesammim	poisoned	M5	tesammim A	tesammim A	tesammim A	tesammim A	tesammim A
ingall	taken	M7	ingall A	ingall A	ingallee A	ingall A	ingall A
irtadd	be returned	M8	irtadd A	irtadd A	irtaddee A	irtadd A	irtadd A
steġall	exploit	M10	steġIll A	steġIll A	steġallee B	steġall B	steġall B
hollow series							
baaʕ	sell	M1	biiʕ A	biiʕ A	biʕ B	baaʕ C	baaʕ C
baat	stay over	M1	baat A	baat A	bit B	baat A	baat A
injaab	be brought	M7	injaab A	injaab A	injib B	injaab A	injaab A
iltaam	blame	M8	il(t)aam A	il(t)aam A	il(t)um B	il(t)aam A	il(t)aam A
stefaad	benefit	M10	stefiid A	stefiid A	stefad B	stefaad C	stefaad C

can be obtained from a mix of different verb measure/series. Should each morphological stem of each series in M1 be included in abstracting the stem space of TLA, then the result will have to include seven different stems cells obtained from all the series, but within each series, there can be no more than four indexed stems. These seven slots are associated with a morphologically conditioned stem in at least one of the verb series in M1. Therefore, each verb lexeme in TLA has a stem space with at least four indexed stems. According to Boyé and Cabredo-Hofherr (2006: 5), 'the maximal stem-space gives the maximally possible number of stem-suppletions.'[11] In TLA, the analyzable forms which can be used as the main criteria for generating the maximal stem-space are represented by the verbal paradigm of the M1 sound series, M1 defective series, M1 double series and M2 defective series.

The initial stem-space in Table 3.11 represents the morphologically motivated instances of stem change that can possibly be found in all verb measures/series of the system including the verbs *klee* 'eat' and *xdee* 'take.' It has eight stem slots and this size is mainly motived by the partial suppletion in the S^i slot that can be found only in the stem inventory of the two high-frequency verbs *klee*

TABLE 3.11 *The initial TLA stem-space*

Subject						3SGM.Obj	
Imperfective		Imperative	Perfective			Perfective	
3SGM	3PLM		1PL	3SGM	3SGF	1PL	3SGM
S1	S2	S^i	S3	S4	S5	S^j	S^k

11 For the purpose of motivating slots in the stem space, Boyé and Cabredo-Hofherr (2006) suggest a number of criteria based on suppletion. According to these criteria, the verb inflectional form can belong to two decomposition types. The first type consists of two subparts (stem + ending) while the second type is one part (form). The former is an analysable stem-suppletion type while the latter is an unanalysable stem which does not have the ending sub-part, but rather is listed as a form suppletion type. Nevertheless, the latter can determine the stem-affix boundaries.

'eat' and *xdee* 'take.' It is important to note that the reason for considering the verbs *klee* and *xdee* in the initial formation of stem-space is based on the account presented by Boyé and Cabredo-Hofherr (2006) according to whom the stem space is derived by analyzable stem-suppletion while unanalyzable forms have to be listed and excluded from the organization of a stem-space. Both *klee* 'eat' and *xdee* 'take' have analysable stem-suppletion forms, including the forms in the imperative, thus, they should have a slot designated in the stem-space. However, enlarging the stem-space based on the stem pattern of only two verbs would depend on 'the regularities one wants to be able to express.' (Montermini & Boyé 2012: 72). Given that this is not a systematic suppletion-pattern in the language, we opt for excluding these verbs from determining the stem-space organization. Consequently, the maximal stem inventory required in the system will be reduced to *potentially* include seven slots for seven stems morphologically motivated.

The seven stem slots are filled by stems that are partial suppletive in independent cases. Each of these stems corresponds to a morphologically induced stem change in one or more verb series of M1 that represents the maximally possible stem-suppletion pattern that can be found in the system (Table 3.13). For instance, S1 can be identified as an indexed stem by the stem change pattern in any series of M1 while S3 is represented by the M1 double and/or hollow series. By contrast, S2 and S5 can be distinguished by the M1 sound series. S^j is listed due to the morphologically induced monophthong elision in the M1 double series whereas S^k has a morphological stem in the defective series (i-vowel final verbs). Therefore, seven stem slots in the stem-space can be sufficient to

TABLE 3.12 *TLA stem-space excluding the verbs* klee *'eat' and* xdee *'take'*

Subject					3SGM.Obj	
Imperfective		Perfective			Perfective	
3SGM	3PLM	1PL	3SGM	3SGF	1PL	3SGM
S1	S2	S3	S4	S5	S^j	S^k

TABLE 3.13 *Examples of verb morphological stems (in greyed cells) which motivate the* TLA *stem-space*

series	verb	Gloss	Subject					3SGM.Obj	
			Imperf			Perf		Perf	
			3SGM	3PLM	1PL	3SGM	3SGF	1PL	3SGM
sound	ṣehad	burn	aṣhid	aṣuhd	ṣehad	ṣehad	ṣuhud	ṣehad	ṣuhud
hollow	šaaf	see	šuuf	šuuf	šuf	šaaf	šaaf	šaaf	šaaf
defective	nfee	exile	infi	ifn	nfee	nfee	nf	nfa	nfa
double	gall	take	gill	gill	gallee	gall	gall	gall	gall
defective	garra	teach	garri	garr	garree	garra	garr	garra	garra
stem space			S1	S2	S3	S4	S5	S^j	S^k

provide the 'overall distribution schema'[12] of TLA morphologically induced stems. It is worth noting that the allomorphic variation displayed by the verb lexemes varies, depending on the verb series/measure rather than a regularity distinction of lexemes.[13]

Table 3.13 shows that the verb inflectional complexity in TLA can be characterised by a stem-space that consists of a set of stems, each of which is connected to a slot in the inflectional paradigm. In the stem slots in the imperfective paradigm, the inflectional forms can be built on the same or distinct stems, depending on the verb series/measure. By contrast, the 1/2SG/PL perfective inflectional forms share the same stem in the stem-space slot in all verbs. The 3SG/PL perfective inflectional forms may select the same or distinct stem slots: the 3SGM slot and 3SGF/PL slot (Table 3.10 and Table 3.13). In fact, the patterns of stem change in most of the non-M1 sound verbs partition the whole paradigm into a set of cells that tend to be filled by the same stem. Since different verbs have different patterns of stem change, consequently not all verbs can have the same indexed stems in the same stem slots. In other words, the stem space slots/indexed stems in different verbs may correspond to the same

12 The term is used by Pirelli and Battista (2000) for representing the notion of stem-space.

13 A regular lexeme is represented by a single stem while a less regular lexeme can be based on more than one suppletive-stem (Bonami & Boyé 2006).

stem base (e.g. M2, M3, M5, M6 sound series) or to four different stem bases (M1 sound series and defective (i-vowel final) verbs) (Table 3.13). Nevertheless, it is important to note that it may not be necessary to postulate a separate slot in the stem-space for the 3SGM form for the reason that although it shows a partial suppletive change in M2, M3, M5, M6 defective verbs and the M1 hollow series, it can be constructed from other already listed stems. Nevertheless, '... storing more than one is harmless' since the overall structure of the model represents the principal parts of lexemes and increasing the size of the stem-space ultimately 'depends on the regularities one wants to be able to express.' (Montermini & Boyé 2012: 72).[14]

In addition, assuming the generalization of a single slot for the 1PL and 3SGM forms based on M1 sound series and the majority of other verb series, representing the slot of S3 for all verb measures and series may lead to an analytical problem in determining the conjugation pattern. The reason for this is that the stem zone[15] coincides with two different stems in certain verb measure/series. For instance, the hollow series has BC or BA stem change (Table 3.10) and defective series of M2/M3 have CD stem change. In other words, the inflectional forms that are built on these indexed stems can vary depending on the verb series. Therefore, in some cases it is important to specify the verb series in the stem distribution (Table 3.14).

Table 3.14 illustrates that filling the slots of the TLA stem-space has to be based on the stem systematic distribution, which can derive a network of implicative relations (principal parts). However, no verb can achieve the maximal theoretical degree of stem pattern complexity displayed in (Table 3.13 and Table 3.14). In fact, in the majority of the verb measures including the sound series (M2, M3, M5, M6 and M10) and double series (M2, M5 and M10), only one morphologically stipulated indexed stem is needed to reconstruct the whole paradigm.

In addition, as Montermini and Boyé (2012: 72–73) pointed out for Italian, the relationship between the stems and/or stems and forms can be derived by a function. The output of a function can be the same as the input (identity

14 The listing of the affixless 3SGM form as an indexed stem means that it has to be treated as an unanalysable form-suppletion in non-defective verbs (e.g. *maat* 'he died'). Boyé and Cabredo-Hofherr (2006) provide a similar analysis for Spanish 2SG imperative stems such as *pon* 'put,' *haz* 'make' which are treated as listed form-suppletion rather than an analysable stem-suppletion (*haz* + Ø).

15 The terms stem zone refers to a cell/slot in the stem-space specified for suppletions. The stem zone coincides with a cell or cells of the paradigm.

TABLE 3.14 *The distribution of* TLA *indexed stems in the stem-space*

Indexed stem	forms built on these stems in different measures/series
S1	a. imperfective/imperative (M1 sound [before Ø.subj or Ø/consonant-initial.obj suffixes], M1 & non-M1 hollow, M1 & M10 double, M1& non-M1defective [before Ø.subj. or any obj. suffixes]). b. imperfective/imperative/perfective (non-M1[a] sound, M2, M5, double)
S2	a. imperfective/imperative (M1 sound [before vowel-initial subj/obj suffixes], M1 & non-M1defective [before vowel-initial subj. suffixes]) b perfective (non-M1defective[before vowel-initial subj. suffixes])
S3	a. 1/2 perfective (non-M1 defective, M1 & M10 double), b. 1/2 perfective & 3SGM (M1 sound and defective)
S4	a. 3 perfective (M1& M10 hollow, M1&M10 double) b. 3SGM perfective (non-M1defective)
S5	perfective 3SGF/3PL (M1 sound, M1 defective)
S^j	perfective 3SGM (M1 double [before consonant-initial obj. suffix])
S^k	perfective 3SGM (M1 & non-M1defective [before vowel-initial obj suffix])

a M7 and M8 are not included.

function). However, the function 'may alter the phonology of the input, this alternation may, but need not be phonologically motivated.' In TLA, the distribution of the indexed stems in the M1 sound and hollow series may not be constrained by a set of default relations expressed by a function, due to the ablaut exhibited by the indexed stems S1, S2, S3, S4, and S5. By contrast, the S^j and S^k stems are based on the ablaut of the final stem vowel in addition to the semi-autonomous morphological characteristics displayed by the unconditioned vowel elision pattern for which it is possible to establish a set of default relations ('stem to stem relations'), connecting the stems in defective series (Montermini & Bonami 2011: 11).

TABLE 3.15 *Stem-to-stem relations in M1 and non-M1 defective verbs*

3SGM.imperf- 3PLM.imperf	1PL.perf- 3SGM.perf	1PL.perf- 3SGF.perf	1PL.perf- 3SGM.obj	3SGM. imperf- 1PL.perf	3SGM.imperf- 3SGM.obj
S1-S2 xv-X	S3-S4 Xee-Xee Xee-Xa	S3-S5 Xee-X	S3-S^k Xee-Xa	S1-S3 VXV-Xee xv-Xee	S1-S^k VXV-Xa
ansa-ans infi-inf bakki-bakk	nsee-nsee nfee-nfee bakkee-bakka	nsee-ns nfee-nf bakkee-bakk	nsee-nsa nfee-nfa bakkee-bakka	ansa-nsee infi-nfee bakki-bakkee	ansa-nsa infi-nfa bakki-bakka

(1) The formalism adopted in (Table 3.15) reflects the following:

a- S1 is related to S2 by truncating the final stem vowel;
b- S3 is related to S4 by identity in M1 or replacing the final vowel with [a] in non-M1;
c- S3 is related to S5 by truncating the final stem vowel;
d- S3 is related to S^k by replacing the final stem vowel [ee] with [a];
e- S1 is related to S3 by truncating the FV and replacing the final vowel with [ee];
f- S1 is related to S^k by truncating FV and replacing the final vowel with [a] or maintaining the final vowel [a].[16]

16 The similar set of functions can also be suggested for the double series stem alternation in M1 and M10.

stem relation in double series	S1-S2 X-X	S3-(S5/S^j). Xee-X.
M1	gill-gill	gallee-gall.
M10	steġill-steġill	steġallee-steġall.

a- S1 is related to S2 by identity;
b- S3 is related to S5/S^j by truncating the final stem vowel.

These relations identify a class of verbs characterized by final stem vowel alternations all of which can be expressed by a function. In the defective series, from S1, it is possible to predict the whole paradigm while other surface stems cannot determine the ablaut in S1. Similarly, as Montermini and Bonami (2011) point out, Italian has cases in which a single stem zone is associated with more than one surface stem each of which is not clearly identified as an indexed stem by lack of a suppletive case.[17] In addition, there are verbs such as *lava* 'to wash,' which show final vowel stem alternations in different stem zones of suppletion. However, these stems themselves are not the reason for providing the different stem slots; hence, they need not be indexed. In other words, each of these stems can be derived by a set of functions expressing the morphologically conditioned alternation of the thematic vowel.

Likewise, in TLA, stem zones S1 to S5 are reflected by the morphologically conditioned ablaut or (partial-suppletion) in M1 sound and hollow series. By contrast, the stem zones S^j and S^k are justified by morphomic stems in defective series motivated by final vowel ablaut, and in double verbs reflected by the absence of the *ee* monophthong (see below) which is morphophonologically conditioned otherwise. If we opt to treat the final stem vowel in the defective verbs in line with the account proposed for the thematic vowel in Romance languages, e.g. Spanish, Italian, and French, (Bonami & Boyé 2002; Boyé & Cabredo-Hofherr 2006), then the defective series (of all measures) and double (M1) series can be derived from the listed stem S1 and/or S3. The relations between S1 and/or S3 and other stems will correspond to final stem-vowel variation or truncation. Similarly, in the double series, the stems can be linked to each other by the identity function or a function specifying the addition of the monophthong *ee*. Following this analysis, the TLA verb monophthong will be similar to the thematic vowel of Romance languages in conditioning regular stem change within that paradigm of regular verbs. In other words, M1 and non-M1 defective and M1 double verbs can be classified as regular verbs that require one indexed stem. This indexed stem can represent suffer stem allomorphic variations located at the site of the monophthong or stem final vowel. Consequently, the TLA stem-space would be reduced to five slots while

17 Likewise, Spanish has four different surface stems for the inflected forms in the stem slot S6 that is identified by a single suppletive case in 'Preterite, Imperfective 1&2 Subjunctive and Future Subjunctive.' In other words, suppletion is the essential criterion for licensing a stem zone in the stem-space while vowel alternations at the edge of the stem do not necessarily require a separate stem slot (Boyé & Cabredo-Hofherr 2006: 8).

TABLE 3.16 *The stem allomorphy of M1 and non-M1 defective series (the XY letters represent the surface allomorphic stem variations of stem A in Table 3.10)*

verb	Gloss	Subject					3SGM.Obj	
		Imperf		Perf			Perf	
		3SGM	3PLM	1PL	3SGM	3SGF	1PL	3SGM
Underlying stems		S1		S3				
nsee	forget	ansa	ans	nsee	nsee	ns	nsa	nsa
nfee	exile	infi	ifn	nfee	nfee	nf	nfa	nfa
bakka	upset	bakki	bakk	bakkee	bakka	bakk	bakka	bakka
naada	invite	naadi	naad	naadee	naada	naad	naada	naada
tewalla	take over	tewalla	tewall	tewallee	tewalla	tewall	tewella	tewella
surface stems		X_i	X_j	Y_i	Y_j	Y_k	Y_l	Y_l

S^j and S^k motivated by the verbs in question[18] may be treated as surface stems morphologically induced by the final stem vowel variation and the default relations between these stem functions. Consider Table 3.16 which shows stem alternations of the defective series in different measures.

In the M1 defective series, the imperfective 3SGM stem in the subject subparadigm can only occur with unsuffixed forms while the imperfective 3PLM stem occurs before a V-initial suffix. However, the stem of the imperfective 3SGM form (or any unsuffixed form) is also used before V/C-initial object suffixes. Likewise, the perfective 1PL/3SGM stem can have the *ee* monophthong in M1, while in non-M1 the corresponding stem has the final vowel *a* which is truncated before V-initial subject suffixes in all measures. By contrast, the monophthong of the 1PL/3SGM stem in M1 alternates with the vowel *a* in the 3SGM.obj-sub-paradigm.

18 It is worth noting that similar to the defective series, in the sound series, S^k is associated with independently motivated ablaut. Nonetheless, it shows a stem-to-stem relation to S4 and can be expressed by the identity function (e.g. S^k *ʕuruf-a* 'he knew him'~ S4 *ʕuruf-it* 'she knew').

Under the stem-space approach, the stem pattern of defectives can be handled by two sub-functions. The first one generates the stem (F_s) while the second realizes the inflected form by combining the stems and the relevant suffixes (F_{+A}). In other words, the relationship between the stem and the forms is expressed by the same formalism as that for the relationship between the stems (Montermini & Bonami 2011). Therefore, through the combination of the sub-functions, the inflectional forms of the defective series (Table 3.16) are morphologically derived from two underlying stems S1 and S3 as in the formalism (2) (adapted from Boyé & Cabredo-Hofherr 2006: 7).

(2)
Underlying Stem ⇒ Surface Stem ⇒ Inflected Form
a. F_S: Underlying Stem ⇒ Surface Stem
b. F_{+A}: Surface Stem ⇒ Surface Stem + A = Inflected Form
c. F_{+A} (FS(Underlying Stem)) = Inflected Form

For the S1 stem, there are two surface stems X_i and X_j respectively for the imperfective 3SGM and 3PLM. In (3) and (4), we give the functions FS1X_i and FS1X_j, realizing the surface stems from the underlying one.

(3)	F_{S1Xi}: S1	⇒	Surface Stem for imperfective 3SGM: S1X_i
	S1	→	S1X_i
		identity	
	F_{S1Xi} (ansa)	=	ansa
	F_{S1Xi} (infi)	=	infi
	F_{S1Xi} (bakki)	=	bakki

(4)	F_{S1Xj}: S1	⇒	Surface Stem for imperfective 3PLM: S1X_j
	S1	→	S1X_j
		truncate final V	
	F_{S1Xj} (ansa)	=	ans
	F_{S1Xj} (infi)	=	inf
	F_{S1Xj} (bakki)	=	bakk

The distinctions between these two surface stems are reflected only by the presence or absence of the final vowel. Likewise, the three different surface stems for the inflected forms in the ablaut change zone (S3) can be derived by the functions FS3Y_i and FS3Y_j in (5), (6) and (7):

(5) F_{S3Yj}: S3 ⇒ Surface Stem for perfective 3SGM: $S3Y_j$
S3 → $S3Y_j$
identity
a. F_{S3Yj} (nsee) = nsee
F_{S3Yj} (nfee) = nfee

b. F_{S3Yj} (bakka) = bakka
F_{S3Yj} (naada) = naada
F_{S3Yj} (tewalla) = tewalla

(6) F_{S3Yk}: S3 ⇒ Surface Stem for perfective 3SGF: $S3Y_k$
S3 → $S3Y_k$
truncate final V
a. F_{S3Yk} (nsee) = ns
F_{S3Yk} (nfee) = nf

b. F_{S3Yk} (bakka) = bakk
F_{S3Yk} (naada) = naad
F_{S3Yk} (tewalla) = tewall

(7) F_{S3Yl}: S3 ⇒ Surface Stem for perfective 3SGM.Obj.3SGM: $S3Y_l$
F_{S3Yl}: S3 ⇒ $S3Y_l$ = S3 with a modification of final V: ee ⇒ a, a⇒ a
a. F_{S3Yl} (nsee) = nsa
F_{S3Yl} (nfee) = nfa

b. F_{S3Yl} (bakka) = bakka
F_{S3Yl} (naada) = naada
F_{S3Yl} (tewalla) = tewalla

This morphological model is based on a realizational analysis according to which functions applied to the stem realize the inflected forms. These functions do not assign features to the stem, but only express a set of features realized by the inflected forms. Therefore, the inflected forms based on S1 and/or S3 can be derived by functions F_N that realize these inflected forms by combining the two sub-functions as represented by the formalism for the constructing the perfective forms of the verb *nfee* 'exile' in (8):

(8)

a.	F_1 (S3)	⇒	perfective 3SGM	e.g. F1 (nfee)	=	*nfee*
b.	F_2 (S3)	⇒	perfective 3SGF	e.g. F2 (nfee)	=	*nf*
c.	F_3 (S3)	⇒	perfective 3SGM.obj.3SGM	e.g. F3 (nfee)	=	*nfa*

where

F_1 (S3)	=	$F_{+Ø}(F_{S3Yj}(S3))$	=	(S3) + Ø
nfee		→		nfee + Ø
		keep the stem, add + Ø		
F_2 (S3)	=	$F_{+it}(F_{S3Yk}(S3))$	=	(S3- ee) + it
nfee		→		nf+ it
		truncate ee, add + it		
F_3 (S3)	=	$F_{+a}(F_{S3Yl}(S3))$	=	(S3 (ee ⇒ a)) + a
nfee		→		nfa + a
		modify final V (ee ⇒ a), add + a[19]		

Clearly, the distinction between surface stems in the defective series of any measure shows that the distinguishing criteria identifying these classes of verbs are the morphologically motivated modification or truncation of stem final vowels. Thus, the relationship between S1 and S3 can be related by assuming a function that deletes the FV and modifies the stem vowel of S1 to the monophthong *ee* in M1 and the low vowel *a* in non-M1. In other words, the stem alternation in this series can be connected by systematic default relations which help reconstruct the whole paradigm of the verb through one and/or two listed stem(s), reducing the TLA stem-space in (Table 3.11) to five zones (Table 3.17).

---

19 The functions F_n for the inflection forms based on the two sub-functions of S3 might possibly be formalised as following:
a. F1 (S1) ⇒ imperfective 3SGM e.g. F1 (infi) = *yinfi.*
b. F2 (S1) ⇒ imperfective 3PLM e.g. F2 (infi) = *yinfu.*
where F_1 (S1) = $F_{+y}(F_{S1Xi}(S1))$ = y + (S1).
infi → y + infi.
keep the stem, add + y.
F_2 (S1) = $F_{+y\text{-}/\text{-}u}(F_{S1Xj}(S1))$ = y + (S1- final V) + u.
infi → y + inf + u.
truncate final V, add + y, add + u.

TABLE 3.17 *TLA stem-space (final version)*

Subject				
Imperfective		Perfective		
3SGM	3PLM	1PL	3SGM	3SGF
S1	S2	S3	S4	S5

It can be clearly seen that based on the stem space framework, the paradigm of TLA verbal inflection can be partitioned into five 'clearly semi-suppletive' zones associated with ablaut. In addition, this approach promotes limiting the stem variation suggested in the zone S^j and S^k by considering them as surface stems derived by functions from an underlying stem. Therefore, these stems are part of a function that constructs the inflected forms based on an underlying stem.

The following section will discuss the local stem relations that structure the stem-space and result in classifying TLA verbs into different ICs based on the stem alternations.

4 Inflection Classes and Stem Relations

The work on stem-spaces in Romance languages (Bonami & Boyé 2002; Montermini & Bonami 2011) distinguishes two types of allomorphy; regular and irregular. In the former, the regular pattern is shown by regular verb paradigms in which the stems (which need not be identical) are linked by regular relations derived from only one indexed stem. The latter type of stem change necessarily involves lexicalised suppletive stems with distinct forms that cannot be determined by another listed stem. Based on the pattern of stem allomorphy, it is possible to distinguish regular and irregular inflection classes. The regular class is the first default class in which verb lexemes show regular stem allomorphy while the irregular class in the non-default class the members of which have unrelated stems (Bonami & Boyé 2006).

Likewise, the stem allomorphy in TLA can distinguish different ICs, which leads to the reclassification of the verb measures and series into different sets of verb classes. The combination of these stem slots represents the conjugation pattern in TLA which can have six different classes across all verb forms, four of which are patterns of the M1 verb. The criteria for different conjugations can be based on the pattern of stem change across all the stem-space slots. If a verb has the stem alteration pattern AA and another verb as AB then these two verbs will classify as two different conjugations (Table 3.18).

Table 3.18 shows that the pattern of stem allomorphy can be manifested in six different ICs (Table 3.19). Class I is the dominant IC in the system and it includes the lexemes in which the *allostems* (Spencer 2012) are derived from a single underlying stem. For instance, the sound series {M2, M3, M5, M6, M10} has regular stem allomorphic relations. Likewise, the defective series of any measure has 'RELATED STEMS' (Boyé & Cabredo-Hofherr 2006) allomorphy based on the default relation which expresses variations located at the site of the final stem vowel. By contrast, the other classes, including the regular verb series, specify suppletive stems, resulting in two to four isolated stems. For example, the M1 sound series has the highest number of unrelated stems. In fact, unlike Romance languages, in TLA, the class with stem alternations is represented by what is traditionally considered the regular verb series of the default (simple/basic) verb measure.

TABLE 3.18 *The conjugation pattern in TLA based on stem alternations*

verb	Gloss	verb series.M	Imperfective		Perfective		
			3SGM	3PLM	1PL	3SGM	3SGF
getal	kill	sound M1	ugtil A	ugutl B	getal C	getal C	gutul D
ʕallim	teach	sound M2	ʕallim A	ʕallm A	ʕallim A	ʕallim A	ʕallm A
insemaʕ	be heard	sound M7	insimaʕ A	insimiʕ B	insemaʕ C	insemaʕ C	insimiʕ B
gall	take	double M1	gill A	gill A	gallee B	gall B	gall B
baaʕ	sell	hollow M1	biiʕ A	biiʕ A	biʕ B	baaʕ C	baaʕ C
baat	stay over	hollow M1	baat A	baat A	bit B	baat A	baat A
Stem space			S1	S2	S3	S4	S5

TABLE 3.19 *The TLA ICs under the stem-space framework*[20]

Class	Imperfective		Perfective			verb measures and series
	3SGM	3PLM	1PL	3SGM	3SGF	
SS	S1	S2	S3	S4	S5	
I	A	A	A	A	A	sound {M2, M3, M5, M6, M10} double {M2, M5, M7, M8} defective {M1, M2, M3, M5, M6, M7, M8, M10}
II	A	A	B	A	A	hollow {M1, M7, M8}
III	A	A	B	B	B	double {M1, M10}
IV	A	B	C	C	B	sound {M7, M8}
V	A	A	B	C	C	hollow {M1, M10}
VI	A	B	C	C	D	sound {M1}

TABLE 3.20 *The set of related stems in Classes II–VI*

verb class	stem relations
Class II	{S3},{S1, S2, S4, S5}
Class III	{S1, S2}, {S3, S4, S5}
Class IV	{S1},{S2, S5},{S3, S4}
Class V	{S3},{S1, S2},{S4, S5}
Class VI	{S1},{S2},{S5}{S3, S4}

Unlike class I, the pattern of stems in class II-V cannot always be defined in terms of one another due to a morphologically conditioned ablaut change which results in one or a set of isolated stems. However, Bonami and Boyé (2002) propose that the distribution of unrelated stems inside the stem-space

20 Specifying a separate slot for 3SGM form does not affect the number of inflection classes in the system. However, it can show further variations or distinctions among the different categories.

is not entirely arbitrary and it can show some degree of organization. Consider the following stem relations in Class II–VI.

The verb categories can be based on two to four unrelated sets of related (identical) stems. Therefore, filling the stem-space is constrained by dependency relations of varying complexity. The default stem is S1 (the stems of the imperfective sub-paradigm) that predicts most of the stem-space. Therefore, similar to French, the TLA stem-space can be constrained by dependency relations (9) that hold among the stem sets in a verb paradigm.

(9) The dependency relation in TLA verb paradigms
{Non-perfective [3SGM IDENTICAL OR PARTIAL_SUPPLETION → 3PLM]} IDENTICAL OR PARTIAL_SUPPLETION → {Perfective [3SGF/PL IDENTICAL OR PARTIAL_SUPPLETION → 3SGM IDENTICAL OR PARTIAL_SUPPLETION →1.2SG/PL]}.

This generalization works for most of the verb measures/series. TLA has a dependency relation between the two stems used in the perfective sub-paradigm. The 3SGF/PL form is either identical with or unrelated to the 3SGM stem; and the 1/2SG/PL is either partially-suppletive or identical with the 3SGM stem. However, in M1 sound, hollow (with stem change pattern 1) and double verbs, all the stems in the perfective sub-paradigm are partial-suppletion of the default stems in the imperfective sub-paradigm. By contrast, in the M1 and non-M1 defective verbs, the dependency between the stems tends to be based on identical relations (Table 3.16). These dependency relations can be based on the inheritance tree structure. For example, M1 double and hollow series have the dependency relations illustrated in (Figures 3.1 and Figure 3.2 respectively).

The TLA dependency account relies on the observation that there are identical and/or ablaut change relationships between the stems of these verb series. The dependency relations within the perfective are as follows: in sound verbs, the 3SGF/PL/perf is a semi-suppletion of the stem zone 3SGM, which has the identical dependency relation with the stem in the 2SG/PL/perf slot. By contrast, in hollow verbs, the stem in the 3SGM/perf slot is identical to that in the 3FSG/PL/perf slot and a suppletion to the 12SG/PL/perf stem zone. In other words, three distinct stems need to be stipulated for hollow verbs as shown in the inheritance tree (Figure 3.2). These patterns of dependency relations provide important implications for the inflectional regularity and ICS in TLA.

In French (Bonami & Boyé 2006), a regular lexeme is generally associated with a uniform stem (e.g. *lav*) while semi-regular verbs show suppletion. Unlike

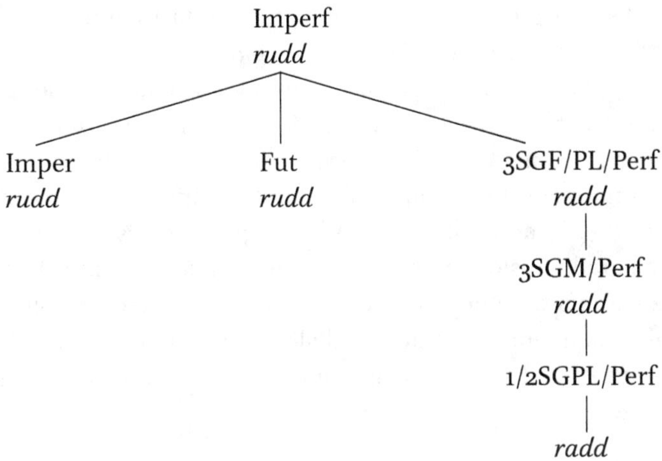

FIGURE 3.1 *The stem dependency tree for the double verb* radd *'return'.*

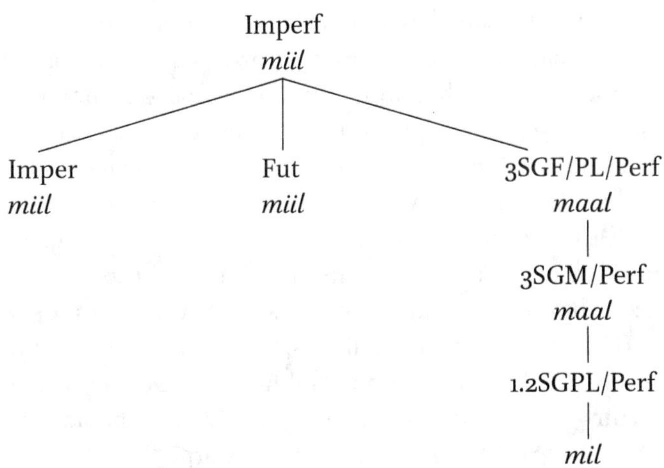

FIGURE 3.2 *The stem dependency tree for the hollow verb maal 'lean'.*

French, in TLA uniformity of the stem choice and the ICs based on the same stem is not a characteristic of the regular sound verbs, which stipulate distinct stems across the sub-paradigms, while some non-M1 verbs use only a single stem for all the inflectional forms. Therefore, regularity in this dialect may not be based on the uniformity of the stem, but rather it seems that the canonical stem has to vary in the verb paradigm of TLA. The fact that it is more regular to have multiple stems than to have just one stem is reminiscent of the situation in English: regular verbs have a present stem and a past (dental) stem: (*walk* ~ *walked*), but some irregular verbs have only one stem (*put* ~ *put*).

In summary a stem-based account for TLA verbs shows that inflectional complexity can be defined by stem sets provided by the stem-space and connected by 'dependency relations' listed by intermediate nodes of the inheritance hierarchies to minimize the amount of redundant information to be stored for a given lexeme. In addition, the TLA stem-space consists of stem slots selected by five paradigm slots based on unrelated indexed stems, thus, the stem-space is constrained by dependency relations of varying complexity. Clearly, using the dependency relations modelled by the inheritance tree in filling the stem-space of verbs avoids stem redundancy in the lexicon and predicts six ICs for TLA.

5 Conclusion

In traditional descriptions, it has been assumed that variation in the shape of the lexical items in Semitic languages is governed by purely phonological rules and that the status of Arabic verb morphology is an autosegmental morpheme-based system. However, in this paper, I have shown that TLA makes extensive use of verb stem variation that cannot be reduced to regularities of sound structure, but which is correlated with morphological rather than phonological factors, hence, the morphomic patterning revealed in TLA conjugation poses serious empirical problems for any morpheme-based account.

Section 2 introduced the morphsyntactic features of TLA, which are expressed by affixation. However, the inflectional exponents in all verbs including the irregular ones do not show allomorphy and instead TLA shows stem alternations distinguishing distinctive cells of the perfective verb paradigms, (3PL/3SG.F from 1/2 and the 3SG.M forms), which has a morphomic distribution. In addition, the TLA verbal paradigm exhibits implicative relations; hence, the imperative and future forms do not necessary need to be listed since they are always predictable from the default imperfective form of any verb even for the highly irregular verbs.

This work provided a stem-based account, showing that inflectional complexity can be defined by stem sets provided by the stem space and connected by 'dependency relations' listed by intermediate nodes of the inheritance hierarchies to minimize the amount of redundant information to be stored for a given lexeme. I proposed the TLA 'stem space' which consists of stem slots selected by five paradigm slots based on unrelated indexed stems filling the stem space, which is constrained by dependency relations of a varying complexity.

Clearly, using the dependency relations modelled by the inheritance tree in filling the stem space of TLA verbs avoids stem redundancy in the lexicon and predicts six ICs for the TLA.

References

Anderson, S.R. 2011. Stress-conditioned allomorphy in Surmiran (Rumantsch). In M. Goldbach, M.-O. Hinzeln, M. Maiden, & J.C. Smith (eds.), *Morphological autonomy: Perspectives from Romance inflectional morphology*, 13–35. Oxford: Oxford University Press.

Aronoff, M. 1994. *Morphology by itself*. Cambridge, MA: MIT Press.

Baerman, M. & Corbett, G. 2012. Stem alternations and multiple exponence. *Word Structure* 5(1). 52–68.

Benmamoun, E. 1999. Arabic Morphology: The central role of the imperfective. *Lingua* 108. 175–201.

Benmamoun, E. 2003. The role of the imperfective template in Arabic morphology. In J. Shimron (ed.), *Language processing and acquisition in languages of Semitic, root-Based, morphology*, 99–114. Amsterdam: John Benjamins.

Bonami, O. & Boyé G. 2002. Suppletion and dependency in inflectional morphology. In Van Eynde F. Hellan, L. & Beerman D. (eds.), *Proceedings of the 8th International Conference on Head-Driven Phrase Structure Grammar* (HPSG), 51–70. Stanford: CSLI Publications.

Bonami, O. & Boyé G. 2006. Deriving inflectional irregularity. In S. Müller (ed.), *Proceedings the 13th International Conference on Head-Driven Phrase Structure Grammar* (HPSG), 361–380. Stanford: CSLI Publications.

Bonami, O. & Boyé G. 2008. Paradigm shape is morphomic in Nepali. Paper presented at *the 13th International Morphological Meeting*, Vienna.

Boyé, G. & Cabredo-Hofherr, P. 2006. The structure of allomorphy in Spanish verbal inflection. *Cuadernos de Linguistica del Istituto Universitario Ortega y Gasset* 13. 9–24.

Camilleri, M. 2014. *The stem in inflectional verbal paradigms in Maltese*. University of Surrey dissertation.

Crysmann, B. & Bonami, O. 2012. Establishing order in type-based realizational morphology. In S. Muller (ed.), *Proceedings of the 19th International Conference on Head-Driven Phrase Structure Grammar, Chungnam National University Daejeon*, 123–143. Stanford, CA: CSLI Publications. http://web.stanford.edu/group/cslipublications/cslipublications/HPSG/2012/crysmann-bonami.pdf (15 August, 2014.)

Darden, B. 1992. The Cairene verb without form classes. In D. Bentari, G.N. Larson & L. A Macleod (eds.), *The joy of grammar: A Festschrift in honor of James D. McCawley*, 11–24. Amsterdam: John Benjamins.

Elfitoury, A. 1976. *A descriptive grammar of Libyan Arabic.* University of Georgetown dissertation.

Gafos, I.A. 2003. Greenberg's asymmetry in Arabic: A consequence of stems in paradigms. *Language* 79. 317–355.

Goldsmith, J. 1976. *Autosegmental phonology.* MIT, York: Garland. Doctoral dissertation

Guerssel, M. & Lowenstamm, J. 1996. Ablaut in Classical Arabic measure I active verbal forms. In J. Lecarme, J. Lowenstamm & Shlonsky, U. (eds.), *Studies in Afro-asiatic Grammar,* 123–134. The Hague: Holland Academic Graphics.

Harrama, A. 1993. *Libyan Arabic morphology: Al-Jabal dialect.* The University of Arizona dissertation.

Harris, Z. 1941. Linguistic structure of Hebrew. *Journal of the American Oriental Society* 61. 143–167.

Heath, J. 1987. *Ablaut and ambiguity: Phonology of a Moroccan Arabic dialect.* New York, Albany: State University of New York Press.

Heath, J. 1997. Moroccan Arabic phonology. In A.S. Kaye & P.T. Deniels (eds.), *Phonologies of Asia and Africa Inc Caucasus,* vol. I, 205–217. Indiana: Winona Lake.

Heath, J. 2003. Arabic derivational ablaut, processing strategies, and consonantal roots. In J. Shimron (ed.), *Language processing and acquisition in languages of Semitic root-based, morphology,* 100–115. Amsterdam: John Benjamins.

Hoberman, R.D. & Aronoff, M. 2003. The verbal morphology of Maltese: From Semitic to Romance. In J. Shimron (ed.), *Language Processing and Language Acquisition in Languages of Semitic, Root-Based Morphology,* 61–78. Amsterdam/Philadelphia: John Benjamins 62. 143–167.

Maiden, M. 2009. From pure phonology to pure morphology the reshaping of the romance verb. *Recherches Linguistiques de Vincennes* 38. 45–82.

Maiden, M. 2011. Morphophonological persistence. In A. Ledgeway, M. Maiden,

J.C. Smith (eds.), *The Cambridge history of the Romance languages,* Cambridge: CUP, 155–215.

McCarthy, J. 1979. *Formal problems in Semitic phonology and morphology.* University of Massachusetts, MIT.

McCarthy, J. 1981. A prosodic theory of nonconcatenative morphology. *Linguistic Inquiry* 12. 373–418.

McCarthy, J. 1986. OCP effects: Gemination and antigemination. *Linguistic Inquiry* 17. 207–263.

McCarthy, J., & Prince, A.S. 1990. Prosodic morphology and templatic morphology. In M. Eid & J.J. McCarthy (eds.), *Perspectives on Arabic Linguistics* 2. 1–54, Amsterdam: Benjamins.

McOmber, M.L. 1995. Morpheme edges and Arabic infixation. In M. Eid (ed.), *Perspectives on Arabic Linguistics,* vol. VII, 173–189. Amsterdam/Philadelphia: John Benjamins.

Mitchell, T.F. 1960. Prominence and syllabification in Arabic. *Bulletin of the School of Oriental and African Studies* 23. 396–389.

Montermini, F. & Bonami, O. 2011. Stem spaces and predictability in verbal inflection. Ms. University of Toulouse and Paris Sorbonne. Retrieved from http://www.llf.cnrs.fr/sites/llf.cnrs.fr/files/biblio/MonterminiBonami2013.pdf (25 July, 2015.)

Montermini, F. & Boyé, G. 2012. Stem relations and inflectional class assignment in Italian. *Word Structure* 5. 69–87.

Morin, Y.-C. 1987. French data and phonological theory. *Linguistics* 25(5). 815–843.

Owens, J. 1984. *A Short reference grammar of Eastern Libyan Arabic*, Wiesbaden: Otto Harrassowitz.

Owens, J. 1980. The syllable as prosody: A reanalysis of syllabification of Eastern Libyan Arabic. *Bulletin of The School of Oriental and African Studies* 43. 277–287.

Pirelli, V. & Battista, M. 2000. The paradigmatic dimension of stem allomorphy in Italian verb inflection. *Rivista di Linguistica* 12(2). 307–379.

Ratcliffe, R. 1997. Prosodic templates in a word-based morphological analysis. In M. Eid, & Ratcliffe, R. (eds.), *Perspectives in Arabic Linguistics X*, 147–171. Amsterdam/Philadelphia: John Benjamins.

Ratcliffe, R. 1998. *The "broken" plural in Arabic and comparative Semitic: Allomorphy and analogy in non-concatenative morphology*. John Benjamins: Amsterdam

Spencer, A. 1991. *Morphological Theory: An introduction to word structure in generative grammar*. Oxford: Wiley-Blackwell.

Spencer, A. 2012. Identifying Stems. *Word Structure* 5. 88–108.

Spencer, A. 2013. Stems and Inheritance. Retrieved from https://www.academia.edu/4379008/Stems_and_inheritance. (7 February, 2014.)

Ussishkin, A. 1999. The inadequacy of the consonantal root: Modern Hebrew denominal verbs and output-output correspondence. *Phonology* 16. 401–442.

Ussishkin, A. 2000. Root-and-pattern morphology without roots or patterns. In M. Hirotani, A. Coetzee, N. Hall, & J.Y. Kim (eds.), *Proceedings of the North East Linguistic Society (NELS) 30*, 655–610. Amherst: GLSA.

Watson, J. 2002. *The phonology and morphology of Arabic*. Oxford, England: Oxford University Press.

Appendix

TABLE 3.21 *An example of the imperfective paradigm of M1 sound, double, hollow and defective verbs*

verb series	sound	double	hollow	defective
MPS	ʕeraf 'know'	karr 'pull'	jaab 'bring'	gree 'read'
1SG	ni-ʕrif	n-kurr	n-jiib	na-gra
1PL	ni-ʕirf-u	n-kurr-u	n-jiib-u	na-gr-u
2SG.M	ti-ʕrif	t-kurr	t-jiib	ta-gra
2SG.F	ti-ʕirf-i	t-kurr-i	t-jiib-i	ta-gr-i
2PL.M	ti-ʕirf-u	t-kurr-u	t-jiib-u	ta-gr-u
2PL.F	ti-ʕirf-in	t-kurr-in	t-jiib-in	ta-gr-in
3SG.M	yi-ʕrif	y-kurr	y-jiib	ya-gra
3SG.F	ti-ʕrif	t-kurr	t-jiib	ta-gra
3PL.M	yi-ʕirf-u	y-kurr-u	y-jiib-u	ya-gr-u
3PL.F	yi-ʕirf-in	y-kurr-in	y-jiib-in	ya-gr-in

TABLE 3.22 *An example of the imperative paradigm of M1 sound, double, hollow and defective verbs*

verb series	sound	double	hollow	defective
MPS	ʕeraf 'know'	karr 'pull'	jaab 'bring'	gree 'read'
2SG.M	i-ʕrif	kurr	jiib	a-gra
2SG.F	i-ʕirf-i	kurr-i	jiib-i	a-gr-i
2PL.M	i-ʕirf-u	kurr-u	jiib-u	a-gr-u
2PL.F	i-ʕirf-in	kurr-in	jiib-in	a-gr-in

TABLE 3.23 *An example of the perfective paradigm of M1 sound, double, hollow and defective verbs*

verb series	sound	double	hollow	defective
MPS	ʕeraf 'know'	karr 'pull'	jaab 'bring'	gree 'read'
1SG	ʕeraf-t	karree-t	jib-t	gree-t
1PL	ʕeraf-na	karree-na	jib-na	gree-na
2SG.M	ʕeraf-t	karree-t	jib-t	gree-t
2SG.F	ʕeraf-ti	karree-ti	jib-ti	gree-ti
2PL.M	ʕeraf-tu	karree-tu	jib-tu	gree-tu
2PL.F	ʕeraf-tin	karree-tin	jib-tin	gree-tin
3SG.M	ʕeraf	karr	jaab	gree
3SG.F	ʕuruf-it	karr-it	jaab-it	gr-it
3PL.M	ʕuruf-u	karr-u	jaab-u	gr-u
3PL.F	ʕuruf-in	karr-in	jaab-in	gr-in

TABLE 3.24 *The inflectional paradigm of M2 sound series dakkir 'remember/remind' and defective series bakka 'upset'*

MPS	dakkir 'remember'			bakka 'upset'		
	Imperf	Imper	Perf	Imperf	Imper	Perf
1SG	n-dakkir		dakkir-t	n-bakki		bakkee-t
1PL	n-dakkr-u		dakkir-na	n-bakk-u		bakkee-na
2SG.M	t-dakkir	dakkir	dakkir-t	t-bakki	bakki	bakkee-t
2SG.F	t-dakkr-i	dakkr-i	dakkir-ti	t-bakk-i	bakk-i	bakkee-ti
2PL.M	t-dakkr-u	dakkr-u	dakkir-tu	t-bakk-u	bakk-u	bakkee-tu
2PL.F	t-dakkr-in	dakkr-in	dakkir-tin	t-bakk-in	bakk-in	bakkee-tin
3SG.M	y-dakkir		dakkir	y-bakki		bakka
3SG.F	t-dakkir		dakkr-it	t-bakki		bakk-it
3PL.M	y-dakkr-u		dakkr-u	y-bakk-u		bakk-u
3PL.F	y-dakkr-in		dakkr-in	y-bakk-in		bakk-in

CHAPTER 4

Person Agreement on Converbs in Udmurt

Ekaterina Georgieva

Udmurt is a Finno-Ugric language spoken in the Republic of Udmurtia, Russian Federation. According to the Russian Census of 2010, Udmurt has about 325,000 speakers (in an ethnic population of 550,000 speakers).[1] Together with the Komi-Zyryan and Komi-Permyak languages, Udmurt belongs to the Permic subgroup of the Finno-Ugric family. Udmurt uses a wide range of non-finite subordinate clauses. Below, I will concentrate on two converbial clause types, namely the ones headed by the suffixes *-ku* and *-toź*. This paper is organized as follows: first, I will briefly present the usage and the most important morphosyntactic properties of the two converbs in question, and more specifically, their argument structure. Then, I will discuss the person agreement used on the converbs (section 2). In section 3, I will examine whether the agreement on the converbs should be considered pronominal or grammatical agreement. The analysis is based on the referentiality of the person agreement markers and the subject-doubling possibilities in the converbial clauses. Section 4 presents a summary of the discussion. The data presented in this paper come mostly from my own fieldwork conducted in 2013 and 2014 and include elicited sentences, grammaticality judgment questionnaires, as well as spontaneous examples.[2] Additionally, I use some examples from existing descriptive studies and corpora.

1 "The Russian Census of 2010", accessed July 07, 2016, http://www.gks.ru/free_doc/new_site/perepis2010/croc/perepis_itogi1612.htm.

2 I am indebted to my Udmurt consultants for their patience and help when discussing the data presented in this paper. I would also like to express my gratitude to the Surrey Morphology Group, to István Kenesei (University of Szeged & Research Institute for Linguistics, Hungarian Academy of Sciences), András Komlósy (Research Institute for Linguistics, Hungarian Academy of Sciences), Zsuzsanna Gécseg (University of Szeged), to my colleagues from the Departments of Finno-Ugric Linguistics at the University of Szeged and the Udmurt State University as well as to the participants of the 16th International Morphology Meeting (IMM 16) for their comments and suggestions on the early versions of this paper. I am also indebted to Alexandra Kellner for proofreading my manuscript. The support of the CAMPUS Hungary scholarship programme for sponsoring my fieldwork trips is gratefully acknowledged.

 | DOI 10.1163/9789004342934_006

1 The Two Converbs: An Introduction

Udmurt uses a wide range of non-finite clauses. It should be mentioned that the existing descriptive studies on Udmurt deal with non-finite *verb forms* rather than with non-finite *clauses*. The Udmurt non-finite verb forms are traditionally divided into three groups: the infinitive, participles, and gerunds/converbs (Perevoščikov 1962, Bartens 2000, Winkler 2001). Udmurt also has deverbal nouns (masdars, or action nominals). These are not included in this classification, but are instead discussed together with the deverbal derivational suffixes. It should be pointed out that there are many homophonous/polysemic forms, which makes the classification of the non-finites more complicated (Perevoščikov 1962, Georgieva 2012).

Generally, the terminology used in the literature on finiteness is not fully consistent since it makes use of the following terms: *non-finite clauses*, (*lexical* and *clausal*) *nominalizations*, *action nominals* (*masdars*, or *deverbal nouns*), *complex/simple event nouns*, and *process/result nominals* (see Comrie & Thomson 1985, Koptjevskaja-Tamm 1993, 2003, Grimshaw 1990, Haspelmath 1996, Alexiadou 2001 et seq.). Moreover, the non-finite predicates heading adverbial non-finite clauses are often called converbs, gerunds, or sometimes, in accordance with the descriptive traditions regarding the Finno-Ugric languages, participles (see Haspelmath 1995 on the typological term *converb* used in this paper).

Putting the terminological issues aside, what is common in both the typological and the generative literature is that the non-finite clauses are considered to display both verbal and nominal properties. This insight is reflected in Koptjevskaja-Tamm's (1993, 2003) typological classification. She proposes that there are different types of nominalizations: Sentential, Possessive-Accusative, Ergative-Possessive, and Nominal (the labels of these nominalization types reflect the various degrees of similarity of the nominalization to finite verbs or to nouns). A similar idea is found in the generative literature, too. It has been proposed that the non-finite clauses/nominalizations are mixed extended projections, i.e. that they contain an extended verbal projection with a nominal layer on top of it (Borsley & Kornfilt 2000, Alexiadou 2001 et seq., among others). However, the degree of verbal and nominal properties that the non-finite clauses display is a matter of cross-linguistic variation. Moreover, sometimes even one and the same language can employ different non-finite clause types (for instance, English has verbal and nominal *-ing* gerunds, see Abney 1987). In this paper, I refer to non-finite clauses as a general term to denote nominalizations headed by a non-finite verb form that cannot be used in root clauses. I focus on two adverbial non-finite clauses that are headed by the two converbs in question. I assume that these are clausal and not lexical nominalizations (in the sense of Comrie & Thomson 1985).

Typically, a distinction is made between the internal and the external syntax of the non-finite clauses. The internal syntax has to do with the retention/loss of the verbal properties that are characteristic of the finite verbs, such as argument structure (including the expression of the arguments), expression of TAM, and agreement markers, negation, valence, and voice distinctions.

With regard to their external syntax, these clauses can be classified as non-finite relative, non-finite adjunct, and non-finite complement clauses. Roughly speaking, the participles correspond to non-finite relative clauses, whereas the clauses headed by converbs function as adjuncts. Complement non-finite clauses are headed by either an infinitive or a deverbal noun (on the typological classification of the non-finite clauses, see Haspelmath 1996 and Ylikoski 2003, among others).

In this paper, I deal with two of the converbs, those formed with the suffixes *-ku* and *-toź*. Below, I will briefly exemplify the usage of the two converbs in question. As far as their external syntax is concerned, both converbs are heads of temporal/aspectual adjunct clauses. However, in this paper, I mainly focus on their internal syntax rather than their external syntax. More specifically, I will deal with the agreement used in these non-finite adjunct clauses.

1.1 *The Usage of the Two Converbs*

The *-ku*-clauses express an event that is simultaneous or at least partly simultaneous with the one from the main clause. They roughly correspond to English 'when'- and 'while'-clauses or to bare *-ing* gerunds, see (1).[3] Typically, these clauses do not express conditional semantics, i.e. they cannot be used as 'if'-clauses. However, these adjunct clauses can denote habitual events, as in the example below.

(1) (Georgieva 2012, adapted from (Perevoščikov 1962: 286))

[*Gožja-ku*]	*predloženi-len*	*pum-az*	*toćka*	*pukt-o.*
write-NF	sentence-GEN	end-3SG.INES	full stop[ACC]	put-PRS.3PL

'When writing one puts a full stop at the end of the sentence.'

3 Throughout the paper, the non-finite clauses are marked in square brackets. Unless otherwise specified, the source language of the examples is Udmurt. The punctuation and orthography of the original examples is not modified, the only exception being the notation of the clitics, see section 2 below.

The converb *-toź*, on the other hand, has three different meanings. Firstly, it can head a 'while'-clause, as in (2). In this case, the converb *-toź* is similar, though not identical, to the converb *-ku*. The difference is that the non-finite clause headed by the converb *-toź* sets a temporal frame in which the event expressed by the main clause takes place (perhaps carrying the presupposition that the event from the main clause takes place only within this temporal frame). In this way, the complex clause is interpreted as delimitative/durative and its duration is specified by the non-finite clause.

(2) (Perevoščikov 1962: 288)

[Vuri-śkį-toź-amį]	*pęrtem*	*kįrdźan-jos-tį*	*kįrdźa-śko-m*	[...]
sew-INTR-NF-1PL	different	song-PL-ACC	sing-PRS-1PL	

'While we are sewing, we sing different songs [...]'

The converb *-toź* can also be used to encode another type of temporal relation: a 'by the time'-clause or an 'until'-clause, as example (3). Again, the complex clause has delimitative/durative semantics, however in this case, the event expressed in the main clause does not happen within the temporal frame set by the non-finite clause, but is either posterior or anterior to it.

(3) (Turku_Izhevsk_Corpus/Invozho/Kenesh/B/24:38)

[So	*bertį-toź]*	*ćuk-ez*	*nokin*	*ug*	*jętįlį.*
3SG	come.home-NF	ribbon-ACC	nobody	NEG.PRS.3SG	touch[SG]

'Nobody touches the ribbon until he (the boy) comes home (from the army).'

The third usage of the converb *-toź* is not temporal: it expresses an 'instead of'- or 'rather than'-clause as shown in example (4).

(4) (Perevoščikov 1962: 288)

[...]	*[otįn*	*ulį-toź],*	*śęd*	*tel'-įn*	*ul-o.*
	there	live-NF	dark	forest-INES	live-FUT[1SG]

'[...] I would rather live in a dark forest instead (of living there).'

These three meanings are mentioned in the grammars of Udmurt (Perevoščikov 1962, Bartens 2000). However, during my fieldwork, I noticed that speakers of Udmurt do not use the converb *-toź* in all of the meanings listed above. For instance, some speakers do not use it for expressing simultaneous events as in example (2), but prefer the converb *-ku* instead. They also differ with respect to person agreement (see section 2 below).

1.2 *On the Argument Structure of the Converbial Clauses*

In this subsection, I would like to make a couple of remarks concerning the argument structure of these non-finite clauses before proceeding to the person agreement used on the converbs.

To begin with, the *-ku* and *-toź* converbs can be formed from any verb stem (e.g. from transitive, unergative, unaccusative, or weather verbs; examples will be provided in the next sections). Moreover, the *-ku* and *-toź* converbs preserve the full argument structure of the base verb. This is why I consider them clausal nominalizations (and not lexical ones). In the case of converbs formed from transitive verbs, the object retains its accusative marking.[4]

Now let us turn to the expression of the subject in these non-finite clauses. Generally speaking, the subject can be either covert or overt. If it is covert, it can receive either an arbitrary interpretation, as in example (1), or it can be interpreted as referential, as in example (4) (in the latter example, the covert subject is coreferential with the 1SG subject of the main clause). Secondly, these converbial clauses can have an overt subject. In example (3), the overt subject is a 3SG personal pronoun; however, the overt subject can be a noun as well. The overt subject bears the nominative case. The nominative case assignment to the subject is again a property shared by the finite and non-finite clauses.[5] The third possibility is to express the subject by virtue of person agreement markers on the converb. These markers indicate the person/number features of the subject, as in example (2).

It should be emphasized that the converbial clauses can have an independent subject, i.e. their subject does not have to be coreferential with the subject of the main clause. In (3), the subject of the converb (the 3SG pronoun *so* 's/he') is not coreferential with the subject of the main clause. The independent reference of the subject can also be expressed through person agreement. Example (5) is a modified version of (2), in which the subject of the converb is different from the one in the main clause. This is indicated by the 2SG person agreement on the converb:

---

4 Note, however, that Udmurt shows differential object marking (see Winkler 2001: 20–21). The object may be morphologically unmarked or morphologically marked (in the latter case it bears the accusative). This holds for both finite and the non-finite clauses.

5 It can be suggested, though, that the overt subject bears default case (which happens to be the nominative). Similar proposals have been made for Modern Turkish (Kornfilt 2003). This has not been proposed in the case of Udmurt, however. Moreover, it is an assumption that needs to be supported by independent evidence. This task falls beyond the scope of the present study, hence for the time being I assume that the overt subject of the non-finite clauses bears the nominative. (Note that the nominative case is not marked explicitly with [NOM] in the glosses.)

(5) (Based on Perevoščikov 1962: 288, elicited)
[*Vuri-śkị-toź-ad*] (*mi*) *pẹrtem* *kịrdźan-jos-tị* *kịrdźa-śko-m* [...]
sew-INTR-NF-2SG (1PL.EXCL) different song-PL-ACC sing-PRS-1PL
'While you are sewing, we sing different songs [...]'

The possibility of expressing an independent subject in Udmurt non-finite clauses is by no means exceptional in the Finno-Ugric and Turkic language families (as well as cross-linguistically). Similar non-finite clause types have been discussed, for instance, in the case of Hungarian (Sárik 1998), Old Hungarian (Dékány 2014), Finnish (Toivonen 2000), Komi-Zyryan, and Meadow Mari (Serdobolskaya et al. 2012) and the Ob-Ugric languages (Nikolaeva 1999, Bíró 2011), as well as in the case of several Turkic languages (e.g. Modern Turkish, Tatar and Kazakh, see Kornfilt 2003, Sahan 2002, Ótott-Kovács 2015, respectively).

2 Person Agreement in Udmurt Non-finite Clauses

In this section, I deal with the person agreement used on the two converbs in question. This section begins with a discussion of terminological issues, then proceeds to the description of the person agreement on converbs in Udmurt. The description is done mainly from a synchronic point of view. This section introduces the main problems regarding the person agreement on converbs in Udmurt, which are then discussed at length in section 3.

The person agreement markers in Udmurt non-finite clauses are called either *personal suffixes* (Riese 1998, Csúcs 1988) or *possessive suffixes* (Winkler 2001, Bartens 2000). As we will see below, these person agreement markers indeed show similarities to the possessive suffixes used on nouns. However, it has been argued that they do not express possession (Perevoščikov 1959, Edygarova 2010), but rather are instances of person agreement. In accordance with these studies, I will use the term *person agreement* instead of *possessive suffixes* throughout this paper. So far, I have been using the term *agreement* in the very general sense of encoding grammatical information that indicates the person/number features of the subject.[6]

6 Similar discussions can be found in the case of other Finno-Ugric languages, too. For instance, Nikolaeva (1999: 33) uses the term *subject agreement affixes* for the Northern Khanty data and mentions that these affixes: "[...] go back to the possessive affixes but differ from them phonologically". Additionally, Laczkó (2000) has argued that the possessive suffixes

It should be emphasized that there is a striking difference between agreement in finite and non-finite clauses; namely, the usage of person agreement markers in converbial clauses is generally not considered obligatory (Perevoščikov 1959, Winkler 2001, Bartens 2000). The usage of person agreement in converbial clauses is rather understudied, and there are not many generalizations made about the person agreement on converbs in the literature. Bartens (2000: 245), for instance, claims that the person agreement is optional if the subject of the non-finite is coreferential with the subject of the main clause. On the other hand, if the subjects are not coreferential the person agreement on the converbs seems obligatory. However, it seems that the situation is more complex. Georgieva (2012) argues that person agreement in the non-finite context appears to correlate with several factors. Firstly, impersonal non-finite clauses do not show agreement (see 3.1. below). Secondly, person agreement can be omitted in habitual clauses. Furthermore, it seems that there are different types of non-finite clauses. My fieldwork experience shows that the three different usages of the *-toź* converb behave in a different way with respect to agreement (the 'instead of'-clauses typically do not show agreement, while the temporal ones typically do). A similar observation has been made in the case of Northern Khanty. Nikolaeva (1999: 49) claims agreement is not used in irrealis non-finite clauses (these are headed by certain postpositions). Clearly, this question deserves further research.

Table 4.1 shows the person agreement markers of the two converbs in question. These Udmurt converbs inflect for two numbers (singular and plural) and three persons (first, second, third), just as finite verbs do. The person agreement is exemplified with the verb base *mi̮n-* 'go,' and the meaning of the relevant converbial forms is 'when/while/until/instead I/you, etc. go'.

TABLE 4.1 *The person agreement on the converbs* -ku *and* -toź

mi̮n- 'go'					
1SG	*mi̮ni̮-ku-m*	*mi̮ni̮-toź-am*	1PL	*mi̮ni̮-ku-mi̮*	*mi̮ni̮-toź-ami̮*
2SG	*mi̮ni̮-ku-d*	*mi̮ni̮-toź-ad*	2PL	*mi̮ni̮-ku-di̮*	*mi̮ni̮-toź-adi̮*
3SG	*mi̮ni̮-ku-z*	*mi̮ni̮-toź-az*	3PL	*mi̮ni̮-ku-zi̮*	*mi̮ni̮-toź-azi̮*

used on the Hungarian *-Ás* deverbal nouns do not express possession (although they are phonologically identical to the possessive suffixes used on the common nouns).

In the case of the converb *-toź*, the person agreement markers are preceded by the vowel *-a* (segmented in this work as part of the person agreement marker). The same vowel is found in the general possessive paradigm, e.g. *gurt* 'village' > *gurt-am* 'village-INES/ILL.1SG,' *gurt-oź-am* 'village-TERM-1SG.' From a synchronic point of view, the *-a-* might be analyzed as the actual exponent of the inessive/illative case in forms like *gurtam* 'in/to my village'. However, it seems implausible that the *-a-* vowel in the terminative case-marked forms is an inessive/illative marker. It is more likely that it used to be a linking vowel. The reason that we find the same linking vowel in the person agreement paradigm of the converb *-toź* is that the converb has grammaticalized from a terminative case marked deverbal noun. The converbial suffix is still segmentable, it contains a *-t-* deverbal noun suffix and a terminative case suffix *-oź*. The grammaticalization of the deverbal noun into a converb was enforced by the fact that the deverbal noun suffix *-t-* has completely lost its productivity (this happened in all of the Permic languages; see Bartens 2000, Ylikoski 2003).

In addition to occurring on the converbs, the same person agreement markers are used on postpositions, pronouns, adverbs, and numerals. The person agreement on the postposition *ul-* 'under' is exemplified in Table 4.2. Note that the linking vowel *-a-* occurs with this postposition, too. The meaning of these forms is 'under me/you, etc.'

Edygarova (2010) argues that the person agreement markers used on the converbs (and on postpositions, pronouns, adverbs and numerals) constitute a separate paradigm, which differs from the possessive paradigm used on the nouns. The two paradigms differ most crucially in the 1SG—the agreement suffix used on the converbs always contains an *-m*, while the 1SG possessive suffix on the nouns is a vowel only (-(*j*)*e* for alienable possession, and *-i̯/-i* for inalienable possession). The possessive suffixes used on nouns are exemplified in Table 4.3 below. In this table, the noun bears the nominative. The same possessive suffixes are used if the noun is in the dative, ablative, genitive,

TABLE 4.2 *The agreement marking of postpositions*

ul- 'under'			
1SG	*ul-am*	1PL	*ul-ami̯*
2SG	*ul-ad*	2PL	*ul-adi̯*
3SG	*ul-az*	3PL	*ul-azi̯*

TABLE 4.3 *The possessive paradigm of common nouns (alienable possession)*

pukon 'chair'			
1SG	*pukon-e*	1PL	*pukon-mi̯*
2SG	*pukon-ed*	2PL	*pukon-di̯*
3SG	*pukon-ez*	3PL	*pukon-zi̯*

TABLE 4.4 *The possessive paradigm of common nouns (in the instrumental case, alienable possession)*

pukon 'chair'			
1SG	*pukon-en-i̯m*	1PL	*pukon-en-i̯mi̯*
2SG	*pukon-en-i̯d*	2PL	*pukon-en-i̯di̯*
3SG	*pukon-en-i̯z*	3PL	*pukon-en-i̯zi̯*

caritive/abessive, adverbial and approximative cases. These case suffixes follow the possessive suffixes.

Interestingly enough, the 1SG *-m* appears in the general possessive paradigm, too. The same set of person agreement markers used on the converbs is used in the possessive paradigm if the noun bears certain case suffixes (accusative, inessive/illative, elative, terminative, instrumental, egressive, prolative and adverbial). These case suffixes precede the person agreement markers (with the exception of the accusative); see the table below.[7] Table 4.4 presents the possessive paradigm of the noun *pukon* 'chair' in the instrumental case (the person agreement markers are preceded by the linking vowel *-i̯-*, segmented here as part of the possessive suffix, the instrumental case suffix is *-en*, and the meaning of the relevant form is 'with my/your, etc. chair').

The fact that the 1SG *-m* suffix appears in the general possessive paradigm is acknowledged by Edygarova, as well (2010: 72). For this reason, she concludes

7 However, some of the case suffixes show variation between the two orders, i.e. can either precede or follow the possessive suffixes, see Edygarova (2010: 109–111).

TABLE 4.5 *Person agreement markers on finite verbs*

kar- 'do'			
1SG	*kar-o*	1PL	*kar-o-m(i̯)*
2SG	*kar-o-d*	2PL	*kar-o-di̯*
3SG	*kar-o-z*	3PL	*kar-o-zi̯*

that these markers can have two functions—either expressing person agreement (on the converbs) or expressing possession (in the possessive paradigm of the nouns). Moreover, it should be noted that the suffixes used on the converbs and the ones used in the possessive paradigm are completely identical in the plural.

So far, I have been comparing the person agreement on the converbs with the possessive suffixes. It is also worth comparing them with the person agreement suffixes used on the finite verbs. Generally, the possessive suffixes and agreement of the finite verbs are very similar in Udmurt. As has been mentioned above, the 1SG possessive suffix is only a vowel, while the 1SG suffix on the converbs is *-m*. The finite verbs also generally lack the element *-m* in 1SG. In Table 4.5 I illustrate the future tense paradigm of the finite verbs in Udmurt (the future tense marker is *-o-*).

However, the 1SG *-m* is used in interrogative clauses expressing suggestion/demand/obligation (Perevoščikov 1962). The verb form *karom* would be normally understood as a first 1PL future, i.e. 'we will do' (as opposed to the 1SG form *karo* 'I will do'). Nevertheless, example (6) is ambiguous: it can mean either 1SG or 1PL.

(6) (own fieldwork 2014, elicited)
Mar *kar-o-m?*
What[ACC] do-FUT-1
'What shall/should I do?' or 'What shall/should we do?'

In sum, it seems that these person agreement markers have a very wide distribution—they are used on converbs, postpositions, pronouns, adverbs, nouns, and occasionally on finite verbs. This raises the question of whether we are dealing with clitics rather than affixes. Below I will present some positive and negative evidence in favour of this analysis.

It is a well-known fact that clitics generally show less sensitivity to their host, whereas affixes typically have restrictions concerning the host they can attach to (Zwicky & Pullum 1983, Fuß 2005). In this respect, the person agreement markers on Udmurt converbs resemble clitics. Several additional criteria have been proposed in order to distinguish between clitics and inflectional affixes. One of them is obligatoriness—namely inflectional affixes are said to be obligatory, while clitics are not. Since person agreement on converbs is generally regarded as optional (Perevoščikov 1962, Winkler 2001), we might see this as an argument in favour of the clitic analysis. However, the same agreement markers do not appear to be optional when used on adverbs and postpositions. Another hallmark of clitics is that they attach to affixes or other clitics, but affixes typically do not attach to clitics (Zwicky & Pullum 1983). The two converbs in question can (marginally) take case suffixes (probably only dative). In this case, the dative case ending follows the person agreement on the converb (see example (17) below). If the person agreement on the converbs can be followed by a dative case marker, the person agreement marker should be considered a suffix and not a clitic (unless we consider the dative case marker to be a clitic, too). Thus, from a synchronic point of view, we find evidence both for the suffix and the clitic analysis of the person agreement markers.[8]

I would now like to take into consideration the discussion presented in one of the few studies on clitics in Udmurt, namely Arkhangelskiy (2014). His paper is based on Beserman Udmurt (one of the Udmurt dialects), but the same clitics discussed in his paper are used in Standard Udmurt (as well as in the other Udmurt dialects). Following Anderson (2005), he adopts the terms *prosodic clitics* and *morphosyntactic clitics*. In his paper, he deals with the former, i.e. the ones whose prosodic form is deficient and hence fail to qualify as prosodic words. He distinguishes clitics from inflectional morphology based on the stress placement. In Beserman Udmurt (and in Udmurt in general), the last syllable is stressed, except for in the imperative and negated forms of finite

8 As for the diachrony, both the person agreement occurring on finite verbs and the possessive suffixes on nouns are assumed to be of the same origin, namely personal pronouns (Raun 1988). Moreover, in Udmurt, the possessive suffixes and verbal agreement markers are strikingly similar. Because of this fact, it might be suggested that there must have originally been only one set of person markers. As Raun (1988) suggests, these must have been enclitics. Moreover, the 1SG form must have been *-m*, since this is the reconstructed Proto-Uralic ending. This ending has subsequently been lost in both the possessive and verbal agreement paradigms (see Bartens 2000, among others). Along these lines, it seems very plausible to consider the person agreement of the converbs to be a remnant of the original person enclitics.

verbs. Arkhangelskiy (2014: 6) argues that: "[w]hen a word is accompanied by one or more clitics, the position of the stress remains the same, all clitics being unstressed." He considers the following prosodic clitics in Beserman Udmurt: *no* 'and' (glossed as =ADD), *ńi* 'already,' *uk* 'after all; focus (predicates),' *na* 'else; still; yet,' *ke* 'if,' *ik* 'the same; focus particle,' *a/wa* 'question marker,' *pe* 'quotative marker,' *pi* 'autocitation marker' and *val* 'retrospective shift.'[9] Most importantly, the person agreement markers are not discussed in his paper despite the fact that they are indeed prosodically deficient. In my view, Arkhangelskiy's (2014) study presupposes that the person agreement markers are inflectional affixes rather than clitics. This might be due to the fact that the existing descriptive studies on Udmurt do not even suggest that these markers could be analyzed as clitics, and they are undoubtedly classified as inflectional affixes. However, there is one fact discussed in the Udmurt dialectological studies that actually supports the clitic analysis. It has been shown that certain dialects (Srednochepeckiy dialect from the Northern dialect group, Kukmorskiy from the (Peripheral) Southern dialect group, and the Beserman dialect) show variation in stress in the case of verbs, converbs, nouns, postpositions, pronouns, i.e. all of the targets the agreement markers attach to (Teplyashina 1970, Kelmakov 1998, Karpova 2005 on these dialects). If agreement is marked on them, either the ultimate or the penultimate syllable of the word form is stressed (of course, this holds only for the syllabic agreement markers, i.e. the ones used in the plural, and, in the case of the converb *-toź*, the singular markers, too). It has been mentioned in the literature that this variation in the stress placement is optional in the dialects in question. However, if the stress falls on the penultimate syllable, the agreement markers would fit Arkhangelskiy's (2014) definition of clitics—firstly because they are not phonologically independent, and secondly, because in the word forms showing these markers stress does not fall on the ultimate syllable.

In this paper, I am less concerned with the phonological properties of the person agreement markers used on converbs (I gloss them as inflectional affixes and not as clitics). What I would like to emphasize is the syntactic distribution of these markers. Their distribution also gives insight into the internal structure of the two Udmurt non-finite clauses in question.

9 It is noteworthy that according to the Udmurt orthography, all of these particles are written as separate words. Following Arkhangelskiy (2014), I gloss all of these particles as clitics except for *val*, which is glossed as AUX.PST (past auxiliary, since this is the past form of the verb *be* which is used in the analytical past continuous tense).

As a preamble to this discussion, I would like to mention one observation made about the two Udmurt converbs in the recent literature. Edygarova (2010: 87) has claimed that the overt subject and person agreement are in complementary distribution. Thus, she considers example (7) to be ungrammatical. According to her, the sentence would be correct if either the person agreement marker or the overt 1SG subject of the converbial clause were left out.

(7) (Edygarova 2010: 87)
* *Ton* [*mon* *škola-je* *mi̮ni̮-ku-m*] *eš-te* *pumita-d.*
2SG 1SG school-ILL go-NF-1SG friend-ACC.2SG meet-PST.2SG
'When I was going to school, you met your friend.'

Hence, we might conclude that we are dealing with pronominal agreement markers. This question will be discussed in greater detail in the following section.

3 Pronominal or Grammatical Agreement?

Pronominal agreement has been a widely discussed topic in the literature on agreement from both theoretical and typological perspectives (Siewierska 1999, 2004, Corbett 2006, Jelinek 1984, Bresnan & Mchombo 1987 and Fuß 2005, among others).

As far as the terminology is concerned, several terms have been used in the literature: *pronominal affixes* (Corbett 2006), *pronominal agreement markers* (Siewierska 2004), *anaphoric agreement* (Bresnan & Mchombo 1987), *clitic pronouns* (Jelinek 1984), and *pronominal clitics* (Fuß 2005). Pronominal agreement has been topic of serious debate in the literature on agreement (see Siewierska 1999, 2004 and Corbett 2006 for an overview from a typological perspective).

Firstly, it is matter of debate whether pronominal agreement markers should be considered an instance of agreement at all. In this study, I do not wish to make any statements about this question and I will use the term *agreement* throughout the paper. Most researchers, especially typologists, propose a tripartite system, for instance *pronominal agreement markers—ambiguous agreement markers—syntactic agreement markers* (Siewierska 2004), *free pronouns—pronominal affixes—pure agreement markers* (Corbett 2006), or *independent pronouns—pronominal affixes—redundant grammatical agreement markers* (Mithun 2003).

Another very serious issue is how to distinguish these three types. Several criteria have been proposed in the literature (Siewierska 1999, 2004, Fuß 2005, Corbett 2006), but nevertheless, none of these tests provide fully satisfying results especially if applied cross-linguistically. The most important test is the possibility of co-occurrence of the person agreement marker with an overt DP (i.e. subject-doubling). The hallmark of pronominal agreement markers is that they cannot co-occur with an overt DP. However, even this test can be problematic since in some languages the co-occurrence may depend on word order or on the type of the overt DP: pronominal or lexical. Additional tests focus on obligatoriness, morphological/phonological form, markedness, etc.

Linguists have also argued on whether pronominal agreement markers are the true arguments of the verb or not. Under one of the two major analyses, it has been assumed that pronominal agreement markers are the true arguments of the verb, while lexical subjects (or objects), if present at all, are adjoined to the pronominal argument or function as a topic/focus of the clause (see for instance Jelinek 1984). This approach is adopted in the LFG framework (Bresnan & Mchombo 1987). Alternatively, one can assume that there is a null subject in addition to the pronominal agreement markers (McCloskey & Hale 1984 for Modern Irish and Welsh).

It has been observed that it is cross-linguistically common for free pronouns to grammaticalize into pronominal agreement markers and later into grammatical agreement markers (see Siewierska 1999, 2004). As noted by Mithun (2003), this is a change in form, distribution, and function of these agreement markers. The change in form involves a loss of phonological independence. As we saw in the previous section, the person agreement markers on converbs in Udmurt are not phonologically independent words since they always appear attached to their host (i.e. to the converb). In the subsections below, I will take a closer look at the change in their distribution and function. The former has to do with the co-occurence of the agreement markers with overt DP arguments (i.e. subject-doubling). The latter involves the loss of referentiality of the agreement markers (Siewierska 1999, Bresnan & Mchombo 1987).

In the following section, I will apply some of the tests that have been discussed in the literature. My claim will be that the person agreement markers on converbs in Udmurt display mixed properties. They seem to be very pronominal in nature with regard to referentiality. On the other hand, I will demonstrate that, at least for some speakers, the overt subject DP and person agreement markers are not in complementary distribution (*pace* Edygarova 2010). Based on my own fieldwork with native speakers of Udmurt, I will show that with respect to this test, there is variation among Udmurt speakers, and that some of the speakers do accept subject-doubling in the converbial clauses. I will suggest that an ongoing change is taking place in the grammar of Udmurt, according to which the former pronominal agreement markers are becoming ambiguous agreement in Siewierska's (2004) terms. However, I will claim that they are at a very early stage of this diachronic development. In order to demonstrate this explicitly, I will draw parallels with other Finno-Ugric languages, such as Finnish (addressing the analysis presented in Toivonen 2000, as well some counterarguments brought up by Huhmarniemi & Brattico 2015), Hungarian (Tóth 2000, 2011), and Northern Khanty (Nikolaeva 1999). These related languages show some similarities regarding the person agreement markers used in non-finite clauses; however, there are also differences. Additionally, I will make use of the typological descriptions of other, unrelated languages (Central Alaskan Yup'ik, Navajo and Bininj Gun-Wok, as presented in Mithun 2003 and Evans 1999, respectively).

3.1 *Referentiality*

One of the proposed criteria for distinguishing pronominal agreement markers from grammatical agreement has to do with the referentiality of these affixes (Corbett 2006: 103, Mithun 2003). By referentiality I mean that these markers refer to a particular individual whom the hearer can correctly identify based on the discourse context. This individual, i.e. the referent, must be part of the hearer's knowledge, moreover, it must be contextually salient (see Chafe 1994: 93–107). It is assumed that there are different referents: some of them are specific individuals (encoded as referential expressions), but others are non-referential (for instance, indefinite pronouns, negative pronouns, content question words, and expletives are non-referential).

In the typological literature on agreement, it has been observed that pronominal agreement markers, like free pronouns but unlike grammatical agreement, tend to be referential (Siewierska 1999, Mithun 2003, Corbett 2006). This means that they refer to a particular individual that is established either by the speech event (in the case of first and second person) or by a lexical DP in the preceding discourse.

With respect to this criterion, I will discuss two cases: weather predicates and impersonal sentences. In both cases, the subject is referentially deficient since it does not have an antecedent. The impersonal constructions discussed in this study resemble English finite clauses with the impersonal pronouns *you* and *they* or English infinitives with an arbitrary null subject (*It's great to be here*).

Let us start with weather predicates. Udmurt has several weather verbs; below, I use the verb *zori̮ni̮* 'to rain.' Finite clauses with weather verbs do not have an overt subject (contrary to their English equivalents, which contain the expletive *it*). Moreover, an overt 3SG free pronoun in these clauses leads to ungrammaticality, cf. (8).

(8) (own fieldwork 2014, elicited)

(*So)	zor-e.
3SG	rain-PRS.3SG

'It is raining.'

Now let us turn to converbial clauses with weather predicates. Similarly to the situation in finite clauses such as (8), converbs formed from weather verbs cannot have an overt 3SG pronominal subject (9a). Crucially, 3SG person agreement on the converb yields to an ill-formed sentence, too, cf. (9b). This means that the person agreement on converbs patterns after the free pronouns in the finite clauses since both are incompatible with the non-referential subject of the weather predicate. As (9c) shows, the only grammatical sentence is when the converb formed from a weather verb does not show agreement.

(9) (own fieldwork 2014, elicited)

a. *

[So	*zori̯-ku]*	*park-e*	*mi̯ni̯-ni̯*	*ẹm*	*bi̯gat-e.*
3SG	rain-NF	park-ILL	go-NF	NEG.PST.1PL	can-PL

Intended: 'While it was raining, we couldn't go to the park.'

b. *

[Zori̯-ku-z]	*park-e*	*mi̯ni̯-ni̯*	*ẹm*	*bi̯gat-e.*
Rain-NF-3SG	park-ILL	go-NF	NEG.PST.1PL	can-PL

Intended: 'While it was raining, we couldn't go to the park.'

c.

[Zori̯-ku]	*park-e*	*mi̯ni̯-ni̯*	*ẹm*	*bi̯gat-e.*
Rain-NF	park-ILL	go-NF	NEG.PST.1PL	can-PL

'While it was raining, we couldn't go to the park.'

Note that in (8), the finite weather verb bears 3SG agreement marking, whereas a converb formed from a weather verb cannot show agreement, cf. (9b). Hence, examples (8) and (9) demonstrate not only the similarities between the free 3SG overt pronouns and the 3SG person agreement on converbs, but also the differences between the latter and the 3SG agreement on finite verbs. To put it in a different way, the person agreement on converbs cannot function as default agreement, while the person agreement on the finite verbs can.

Now let us turn to the impersonal clauses. Firstly, I will present the finite impersonal clauses, and then I will proceed with the non-finite ones. In finite impersonal clauses, the verb shows 3PL agreement, however in order to obtain the arbitrary reading, one must use a null pronoun as in example (10a). Sentence (10b) proves that overt 3PL pronouns can never have an arbitrary interpretation (the second translation line in the examples shows the intended arbitrary reading which is possible in (10a) but impossible in (10b)).

(10) (own fieldwork 2014, elicited)

a. *Korka* *pukt-o-zi̯.*
house[ACC] build-FUT-3PL
'They will build a house.'
'Someone will build a house.'

b. *Soos* *korka* *pukt-o-zi̯.*
3PL house[ACC] build-FUT-3PL
'They will build a house.'
* 'Someone will build a house.'

If we turn back to the converbial clauses, we find the same pattern. In order to obtain the arbitrary reading, the agreement on the converb must be omitted (and the subject must be null), cf. example (1) repeated below in (11a). On the other hand, the 3PL agreement on the converbs behaves similarly to the 3PL overt pronoun: it cannot have an arbitrary interpretation (11b). It can only refer to a particular group of referents, established in the discourse, for instance the students in the class.

(11) (Georgieva 2012)

a. [*Gožja-ku*] *predloženi-len* *pum-az* *toćka* *pukt-o.*[10]
write-NF sentence-GEN end-3SG.INES full stop[ACC] put-PRS.3PL
'When writing one puts a full stop at the end of the sentence.'

b. [*Gožja-ku-zi̯*] *predloženi-len* *pum-az* *toćka*
write-NF-3PL sentence-GEN end-3SG.INES full stop[ACC]
pukt-o.
put-PRS.3PL
'When writing they put a full stop at the end of the sentence.' (referential only)

Note again that the finite verbs in (10a, b) as well as in (11a, b) do show 3PL agreement, which seems to be perfectly compatible with non-referential readings. However, in the converbial clauses, the 3PL agreement can have a referential reading only. This example again demonstrates the differences between the

10 Example (11a) was adapted from Perevoščikov (1962: 286).

3PL person agreement on the converbs and the 3PL agreement on the finite verbs on the one hand, and the similarities between the person agreement on the converbs and the free pronouns, on the other.

Northern Khanty seems to be similar to Udmurt with respect to the non-finite clauses with generic/indefinite null subjects. Nikolaeva (1999: 49) shows that in these cases, there is no agreement on the non-finite predicate.

Additionally, I would like to draw a parallel with Hungarian non-finite clauses. Hungarian uses infinitives that can take person agreement. These are referred to as *inflected infinitives* in the literature (see Tóth 2000, 2011 among others). The person agreement markers on the infinitives are called possessive suffixes. The Hungarian inflected infinitives partly resemble the Udmurt converbial clauses showing person agreement with respect to the possibilities of referential/non-referential readings. Tóth (2011) demonstrates that Hungarian 3PL inflected infinitives cannot have arbitrary interpretation, whereas 3PL finite verbs can, cf. (12a, b).

(12) (Tóth 2011: 229, ex. 38a, b)[11]

a.

Kínos	*eb-ben*	*a*	*bolt-ban*	*készpénz-zel*	*fizet-ni-ük.*
Embarrassing	this-INES	DET	shop-INES	cash-INS	pay-INF-3PL

'It is awkward (for them) to pay in cash in this shop.' (arbitrary reading impossible)

b.

Kínos,	*ha*	*eb-ben*	*a*	*bolt-ban*	*készpénz-zel*	*fizet-nek.*
Embarrassing	if	this-INES	DET	shop-INES	cash-INS	pay-PRS.3PL

'It is awkward if they pay in cash in this shop.' (arbitrary reading possible)

However, there is one important difference between the Hungarian inflected infinitives and the Udmurt converbs.[12] Tóth (2000) observes that weather verbs are unlike impersonal sentences since they can occur grammatically with inflected infinitives, see (13). Moreover, omission of the 3SG person agreement leads to ungrammaticality.

(13) Hungarian (Tóth 2000: Ch. 3, ex. (17a))[13]

Március-ban	*már*	*nem*	*kellene*	*havaz-ni-a.*
March-INES	already	not	should	snow-INF-3SG

'In March it should not snow already.'

11 The glossing of the original examples has been slightly modified.

12 I would like to thank Zsuzsanna Gécseg for bringing this fact to my attention.

13 The glossing of the original example has been slightly modified.

In order to account for this fact, Tóth (2000) proposes that the agreement on the infinitives can agree with a quasi-argument like the one of the weather verbs. Tóth (2011) states that the 3PL/3SG marking on the infinitives is always referential (it requires the presence of a referential null subject), whereas the 3PL/3SG agreement on finite verbs can function as default agreement licensing a non-referential null subject (Tóth 2011: 231). She motivates her analysis also with the fact that in Hungarian, finite verbs and infinitives use different 3PL agreement suffixes (Tóth 2011: 232). She also argues that the possessive 3PL suffix has only a referential interpretation, too.

As I have argued in section 2 above, in Udmurt, it is quite difficult to find morphological evidence since the suffixes of the finite verbs and the converbs are very similar. Moreover, they are presumably of the same origin. However, it is obvious that there is a difference between finite and non-finite clauses with respect to the referentiality of the person agreement markers. The 3SG/3PL agreement in finite clauses can have a non-referential reading (as long as the subject is null), whereas the 3SG/3PL agreement in converbial clauses can have a referential interpretation only. This question will be addressed in section 3.3.

Now I would like to draw a parallel with another Finno-Ugric language, namely Finnish (as discussed in Toivonen 2000). Finnish has possessive suffixes that are used not only in noun phrases, but also in non-finite clauses (similarly to the Hungarian inflected infinitives). Toivonen (2000) claims that the possessive suffixes can co-occur with all and only human possessive pronouns within the noun phrase. In example (14), on the other hand, the pronoun *sen* is a non-human (non-personal) pronoun. She argues that the sentence would be grammatical if the 3SG possessive suffix *-nsA* were omitted.

(14) Finnish (Toivonen 2000: 584, ex. (5b))[14]

*	*Minä*	*annan*	*koira-lle*	*sen*	*ruokaa-nsa.*
	I	give	dog-ALL	it.GEN	food-3SG

Intended: 'I give the dog its food.'

Hence, it seems that Finnish possessive suffixes also show sensitivity to the possessor they agree with. Firstly, it has to be a pronoun, and secondly, it must [+human]. This is, of course, slightly different from the situation in Udmurt and Hungarian, but it nevertheless shows that the possessive suffixes do show sensitivity to the referentiality or animacy of the pronoun they agree with.

Interestingly enough, the 3SG Finnish possessive suffix can refer to non-human pronouns (and to lexical possessor as well) as long as they are not

14 The glossing of the original example has been slightly modified.

contained within the same noun phrase, cf. (15) below. This example is perfectly grammatical although the 3SG possessive suffix *-nsA* refers to a non-human pronoun, namely *se*, which is the subject of the clause.

(15) Finnish (Toivonen 2000: 584, ex. (8))[15]
Se heiluttaa häntää-nsä.
it wiggles tail-3SG
'It$_i$ wiggles [*sic* wags] its$_i$ tail.'

As far as referentiality is concerned, I would like to make some remarks from a typological perspective. First of all, Corbett (2006) has argued that we are dealing with a scalar phenomenon. Evidence from other languages has shown quite a wide range of differences in terms of the referentiality of person agreement markers. Evans (1999) has claimed that non-referential uses of a marker are evidence for its status as an agreement marker (for instance, the English 3SG *-s* is used with indefinite and negative subjects). Based on evidence from the Bininj Gun-Wok language (spoken in Northern Australia), he argues that if the markers are obligatory, they are unable to express such contrasts as referential/non-referential and definite/indefinite. On the other hand, Mithun (2003) demonstrates that the pronominal agreement markers in Central Alaskan Yup'ik (Eskimo-Aleut language family) and Navajo (an Athabaskan language) are obligatory, albeit referential. Interestingly enough, she mentions that there is only one context in which the pronominal agreement markers are used non-referentially, namely in the case of weather predicates. This is a very interesting parallel with the Hungarian inflected infinitives, for example. Conversely, it demonstrates that the person agreement markers on Udmurt converbs are even more referential than Hungarian inflected infinitives or Central Alaskan Yup'ik pronominal agreement markers, respectively.

3.2 *Subject-doubling*

The most important criterion for distinguishing pronominal from grammatical agreement is the possibility of subject-doubling—in other words, the co-occurrence of the agreement markers with a DP argument. Based on the data presented in Edygarova (2010), as in example (7) above, we might conclude that the person agreement marking on the Udmurt converbs cannot co-occur with an overt DP, and hence, it is an instance of pronominal agreement. However, there are some counterexamples. Example (16) is from the Northern dialects (Yar district). Example (17) comes from a book of Udmurt

15 The glossing of the original example has been slightly modified.

folk tales; the example is cited in Perevoščikov (1959) and clearly exemplifies the Southern dialects, since the converb suffix is *-kį/-kə̂* and not *-ku* as in standard Udmurt. During my own fieldwork, I have observed very many examples with subject-doubling, for instance, (18) and (19), which represent my consultants' spontaneous speech. Example (18) was produced by a speaker from Igra district (Central-Northern dialect group), whereas example (19) comes from the Kiyasa district (Southern dialect group).

(16) (Karpova 2005: 408)

Wal'l'o,	*[mon*	*bįźį-ku-m],*	*Bud'onnoj*	*kolhoz*
long.time.ago	1SG	get.married-NF-1SG	PN	kolkhoz
nazvatsa	*kari-śkį-l-i-z.*			
called	do-INTR-FREQ-PST-3SG			

'A long time ago, when I got married, the kolkhoz was (still) called Budjonny (a Soviet military commander).'

(17) (Perevoščikov 1959: 66)

[Kuźo-jez	*bertį-kį-z-lį]*	*duriśkiś*	*uno*	*końdon*
master-3SG	come.back-NF-3SG-DAT	smith	many	money[ACC]
daśa-m.				
make-EVID[3SG]				

'By the time his master came back, the smith had made a lot of coins.'

(18) (own fieldwork 2013, spontaneous)

[Mon	*škola-įn*	*dįšetskį-ku-m],*	*tatįn*	*nomįr=no*	*ej*
1SG	school-INES	study-NF-1SG	here	nothing=ADD	NEG.PST
val.					
AUX.PST					

'When I was at school, there wasn't anything here.'

(19) (own fieldwork 2013, spontaneous)

Vorgoron=no…	*[pįjaš*	*ćuk-ez*	*šukkį-ku-z]*	*kunokalįk-jos*
man=ADD	boy	ribbon-ACC	hammer-NF-3SG	guest-PL
kįrdźa-lo	*kįrdźan*	[…]		
sing-PRS.3PL	song[ACC]			

'While the boy is hammering the ribbon (into the beam), the guests are singing a song.'[16]

16 The sentence begins with a false start. The dots after the first word indicate a pause.

I tested the grammaticality of subject-doubling with two questionnaires. The first one was conducted in 2012–2013 and the second was filled out in 2014.

The first questionnaire was filled out online by ten native speakers of Udmurt in October-November, 2012. During my fieldwork in July, 2013, I worked on the same questionnaire with seventeen consultants, so altogether the questionnaire was filled out by twenty-seven native speakers of Udmurt. During this fieldwork trip, the examples from the questionnaire were discussed with the consultants within more general, semi-structured interviews. In this format, I had a chance not only to collect the grammaticality judgments regarding the sentences, but also to ask additional questions about them (preferences, dialectal variation, nuances in the usage, etc.). The speakers came from the following districts of the Republic of Udmurtia: Alnash, Balezino, Debes, Igra, Grakh, Yakshur-Bodya, Kez, Kiyasa, Mozhga, Pichi Purga, and Sharkan. The informants' origin was a very important parameter, since the main goal was to cover the main Udmurt dialect groups (Northern, Central and Southern dialect groups). According to their age, the speakers were divided into three groups: 20–30, 30–40, and 40–50 years old. All of the consultants are bilingual (in Udmurt and Russian), but Udmurt-dominant native speakers, all living in the Republic of Udmurtia.

In the questionnaire, I included sentences containing three types of structures. First of all, the questionnaire contained a converbial clause in which the person agreement was not marked; see (20a) below. The second example was the subject-doubling type, i.e. a converbial clause with an overt subject in the nominative and agreement marking, cf. (20b). The third variant was a converbial clause with agreement marking and an overt subject in the genitive, as in (20c). The last type was included in the questionnaire in order to compare the syntax of the converbial clauses with the structure of the general possessive phrase. It is very common cross-linguistically that non-finite clauses look similar to possessive phrases; see Koptjevskaja-Tamm (1993, 2003). Moreover, genitive-marked subjects of non-finite clauses are found in other Finno-Ugric languages, for instances in the closely related Komi languages (Bartens 2000). In Udmurt, possessors occur in the genitive case and the possessee inflects with possessive suffixes, hence example (20c) has a similar structure: the subject is genitive case-marked and there is a person agreement marker on the converb. However, my hypothesis was that in Udmurt, the genitive-marked subjects would turn out to be ungrammatical or very marginal since I had not encountered such examples in corpora. The consultants were asked to make judgments about the acceptability of the examples (sentences were assessed from 1 to 5 according to their grammaticality). In all of the examples, the converbial clause had an independent subject, i.e. its subject was not coreferential with the subject of the main clause.

Additionally, the goal was to examine whether converbs formed from transitive and intransitive verbs behave similarly with regard to subject-doubling. For this reason, I included examples with the following verbs: *mịnịnị* 'to go,' *uśịnị* 'to fall down' and *korka puktịnị* 'to build a house.' Another parameter tested in the questionnaire was the type of overt DP subject. More specifically, I wanted to check if there is difference between pronominal and lexical subjects with respect to subject-doubling. The examples in the questionnaire had three types of overt subjects: a 1SG pronoun, a 3SG pronoun, and a proper noun.

All in all, the examples from the questionnaire looked like (20a, b, c). In this example, the subject of the converbial clause is a proper noun and the converb is formed from an intransitive (unaccusative) verb.

(20) Example from the questionnaire (own fieldwork 2013)

a. [*Koľa* *uśị-ku*] *Liza* *anaj-ez-lị* *žingịrt-e* *val.*
PN fall-NF PN mother-3SG-DAT call-PRS.3SG AUX.PST

b. [*Koľa* *uśị-ku-z*] *Liza* *anaj-ez-lị* *žingịrt-e* *val.*
PN fall-NF-3SG PN mother-3SG-DAT call-PRS.3SG AUX.PST

c. [*Koľa-len* *uśị-ku-z*] *Liza* *anaj-ez-lị* *žingịrt-e* *val.*
PN-GEN fall-NF-3SG PN mother-3SG-DAT call-PRS.3SG AUX.PST
'When Kolya fell down, Liza was calling her mom/was talking to her mom on the phone.'

Below I will summarize the results of the questionnaire regarding the three structures exemplified in (20a, b, c).

The hypothesis that the genitive-marked subjects are ungrammatical seems to be borne out by the answers of the consultants. In the case of the converb *-ku*, examples like (20c) were rejected by all of my consultants. In the case of the converb *-toź*, a few of my consultants accepted some of the examples or considered them possible in another dialect (but not in their own). However, the speakers' answers did not follow any consistent pattern. Moreover, I have not encountered examples showing pattern (20c) in corpora or written texts, so I consider the genitive-marked subject ungrammatical in the case of both converbs.

As far as examples with subject-doubling are concerned (see (20b)), there is variation among the speakers. My results show that there are two varieties, which I will label variety A and variety B. In variety A, subject-doubling yields ungrammaticality (this is the variety represented by Edygarova 2010). However, speakers of variety B do allow subject-doubling. There is a tendency for subject-doubling with overt pronouns to be more acceptable than with lexical DPs (for some speakers, subject-doubling is grammatical with pronouns only). This holds especially for the 1SG pronoun. The results of the questionnaire show that the valence class of the verb (transitive or intransitive) does not play a role.

Now I will briefly discuss examples like (20a). According to the results of the questionnaire, they are grammatical with lexical subjects. Speakers of variety B, i.e. the ones who accept both (20a) and (20b), sometimes find differences in the meaning of the two variants. These differences are rather poorly understood, and hence, they fall beyond the goals of the present study. What I would like to point out is that examples like (20a) are less acceptable with pronominal subjects. Speakers of variety B correct them with examples like (20b), while speakers of variety A prefer to have the person agreement marked on the converb (but without an overt subject in the converbial clause, since they do not allow subject-doubling).

As far as the geographic distribution of the two varieties is concerned, subject-doubling is more widespread among speakers of the Northern dialects (Kez, Balezino, Igra districts). My results demonstrate that there are some "islands" where agreement on the converbs is strictly pronominal (Mozhga and Sharkan districts), i.e. the Central dialects. Speakers of the Southern dialects (Alnash, Kiyasa and Grakh districts) display variation—some of them do accept examples with subject-doubling, while others do not. Hence, I tend to conclude that the variation can be attributed to differences between the Udmurt dialects.

In the pilot study conducted in 2012 and 2013, I tested subject-doubling only with animate subjects, assuming that subject-doubling would be possible with animate subjects if possible at all. For my fieldwork in 2014, I compiled another questionnaire in which I tested subject-doubling with different types of subjects. The goal was to show whether animacy plays a role, and hence, I included examples with both animate and inanimate subjects. I also tested subject-doubling with non-referential subjects, such as indefinite, quantified subjects/bare quantifiers and wh-subjects. Below, I will summarize the results of this questionnaire.

As far as animacy is concerned, the results of my questionnaire show that subject-doubling with an inanimate subject seems to yield ill-formed results (21)–(22). These examples would be perfectly grammatical if the person agreement on the converb were omitted.

(21) (own fieldwork 2014, elicited)

Tunne	*mon*	*[šundi̯*	*pukśi̯-toź-(*az)]*	*uža-j.*
Today	1SG	sun	set-NF-3SG	work-PST[1SG]

'Today I worked until sunset.'

(22) (own fieldwork 2014, elicited)

[Voina	*kutski̯-ku-(*z)]*	*mone*	*armij-e*	*baśt-i-zi̯.*
War	start-NF-3SG	1SG.ACC	army-ILL	take-PST-3PL

'When the war started, I was taken to the army.'

However, I have attested subject-doubling with an inanimate subject, see (23) below.

(23) (own fieldwork 2014, spontaneous)

A	*[ki̯rdźan-ed*	*pum-az*	*vui̯-ku-z]*	*mon*
And/but	song-2SG	end-3SG.ILL	come-NF-3SG	1SG

ki̯l-jos-se	*ki̯li-śko*	*i*	*šedi-śko,*	*ćto*	*so*
word-PL-3SG.ACC	hear-PRS[1SG]	and	feel-PRS[1SG]	that	3SG

udmurt	*ki̯l-i̯n*	*evel.*
Udmurt	language-INES	NEG.AUX.PRS

'But (then) when your song comes to an end, I hear its words (lyrics) and realise that it's not in Udmurt.'

At this point, I would like to turn back to the discussion of the Finnish possessive suffixes. Toivonen's (2000) observation was that the 3SG possessive suffix cannot agree with a non-human pronoun within the same noun phrase; however, it can refer to a non-human pronouns used within the same clause. In the light of this observation, I also included in the questionnaire an example in which the person agreement marker is not doubled by an overt DP, but it clearly refers to an inanimate referent (the referent was provided in a context sentence, see below). In example (24), the 3SG agreement suffix refers to the noun *šundi̯* 'sun' mentioned in the context sentence. The consultants were given example (24a) but some of them corrected it to (24b).

(24) (own fieldwork 2014, elicited)
Context: *Tunne šundi̯ 7 ćase pukśiz.*
'Today the sun set at 7 o'clock.'

a. ? [*Pukśi̯-toź-az*] *mon* *azbar-i̯n* *uža-j.*
set-NF-3SG 1SG garden-INES work-PST[1SG]

b. [*Šundi̯* *pukśi̯-toź*] *mon* *azbar-i̯n* *uža-j.*
sun set-NF 1SG garden-INES work-PST[1SG]
'I worked in the garden until sunset.'

The data are not very clear. First of all, based on the questionnaire, it seems that the person agreement on converbs: a) cannot be doubled by an overt inanimate DP, see (21) and (22), and b) cannot even refer to an inanimate DP that is available in the context (24a). However, I have attested examples showing subject-doubling with inanimate subjects, cf. (23). I have also observed examples in which person agreement markers refer to inanimate referents (these examples come from the Turku-Izhevsk Corpus). However, they are quite rare. In my opinion, the reason for this is that the person agreement markers generally pick up referents that are established in the discourse, or more precisely, referents that are salient in the discourse. For instance, example (23) was uttered in a context where the main topic of the conversation was the song. On the other hand, the inanimate subjects from the sentences used in the questionnaire, especially given with very little context, cannot be interpreted as salient in the discourse. Hence, I will conclude that the person agreement markers can refer to inanimate referents (just like in Finnish), and presumably, can also be doubled by an inanimate DP (for the speakers of variety B, of course).

Similarly, Nikolaeva (1999: 48–49) has argued that the usage of person agreement in Northern Khanty non-finite clauses is determined by information structure. She demonstrates that the subject of the non-finite clause must be topicalized in order to trigger agreement on the non-finite predicate. By topicalization, she means instances where the subject: "[...] occurs in the previous discourse and is under discussion at the time of the utterance in question" (Nikolaeva 1999: 48).

Two further tests have been suggested in the literature. Both of these show not only whether subject-doubling is possible, but also whether it shows sensitivity to the referentiality of the subject.

The first test is subject-doubling with indefinite DPs. It seems that in this case, subject-doubling leads to grammatical results for those speakers who allow subject-doubling at all (25). Note that in this sentence, the subject is [+animate].

(25) (own fieldwork 2014, elicited)
% [*Ki̮čeke* *ad'ami* *korka-ze* *pukti̮-ku-z*]
Some.kind.of person house-3SG.ACC build-NF-3SG
buskel-jos-i̮z *so-li̮* *veme* *leśt-o.*
neighbour-PL-3SG 3SG-DAT barn.raise[ACC] do-PRS.3PL
'When someone is building a house, their neighbours help them.'

The second test is subject-doubling with a DP taking quantifiers such as *every*. Again, subject-doubling is accepted, see (26).

(26) (own fieldwork 2014, elicited)
% [*Kot'kud(iz)* *pijaš* *ćuk-ez* *šukki̮-ku-z*] *kuno-os*
Every boy ribbon-ACC hammer-NF-3SG guest-PL
ki̮rdźan *ki̮rdźa-lo.*
song[ACC] sing-PRS.3PL
'While every boy is hammering in the ribbon (into the beam), the guests are singing a song.'

However, my results show that subject-doubling with a bare quantifier such as *kot'kin* 'everybody' is, in the best case, odd, see (27).[17] My consultants mentioned that this example was infelicitous not only because of the subject-doubling, but also because of the quantifier in the converbial clause. My explanation for this is that the quantifier has a non-referential meaning. Hence, the meaning of the converbial clause would be: 'when they (non-referential) hammer the ribbon/ when the ribbon is being hammered.' In this way, it is similar to the meaning of the impersonal sentences presented in 3.1., which are generally preferred in non-referential contexts.

17 Udmurt has another quantifier, *vańzi̮* 'all.' However, this one differs from *kot'kin* 'everybody' in many respects. First of all, *vańzi̮* is morphologically complex: *vań* 'all' and *-zi̮* 3PL person agreement marker (3PL possessive suffix). Moreover, the sentences with the quantifier *vańzi̮* show 3PL agreement on the verb. Additionally, there might be other differences between *vańzi̮* and *kot'kin*, for instance in terms of distributive/collective readings. Quantifiers have not been investigated in depth in the case of Udmurt, and thus I leave this question open for further research.

(27) (own fieldwork 2014, elicited)

?? [*Koťkin*	*ćuk-ez*	*šukki̮-ku-z*]	*kuno-os*	*ki̮rdźan*
Everybody	ribbon-ACC	hammer-NF-3SG	guest-PL	song[ACC]

ki̮rdźa-lo.
sing-PRS.3PL

Intended: 'While everybody is hammering in the ribbon (into the beam), the guests are singing a song.'

Another type of non-referential subjects are wh-words. The problem is that this test is not really applicable in Udmurt, since the converbial clauses do not allow for wh-movement. In the best case, we can formulate an echo question, as in (28), in which the wh-word stays in-situ, but echo questions probably do not satisfy the test.

(28) (own fieldwork 2014, elicited)
Context: *Pijaš korkaze pukti̮kuz buskeljosi̮z soli̮ veme leśto.*
'When a/the boy is building his house, his neighbours help him.'

% *Kin*	*korka-ze*	*pukti̮-ku-z?*
Who	house-3SG.ACC	build-NF-3SG

'Building his house who?'

Table 4.6 summarises the subject-doubling possibilities with different types of subjects based on the results from the questionnaire filled out in 2014.

An additional remark must be made regarding the markedness of the converbial clauses showing subject-doubling. It seems that subject-doubled sentences are perceived as more emphatic by the consultants or have a contrastive (focus) reading. However, further research is needed in order to demonstrate the exact factors determining the usage of a converbial clause with or without subject-doubling, especially in non-elicited texts.

TABLE 4.6 *Subject-doubling with different subject types*

	1SG pronoun	3SG pronoun	Proper noun	Indefinite DP	Quantified DP	Bare quantifier	Inanimate DP	Wh-subject
Variety A	no	no	no	no	no	no	no	not applicable
Variety B	yes	yes	yes	yes	yes	??	??	not applicable

3.3 *Explaining the Data*

In the previous sections, I demonstrated that the person agreement on the converbs in Udmurt is very pronominal in nature with regard to referentiality. The subject-doubling criterion, however, shows that there is variation among the speakers. I interpret this as a morphosyntactic change, according to which the originally pronominal agreement is shifting into ambiguous agreement in Siewierska's (2004) terms (i.e. agreement with pro-drop). Evidence in favour of this is the possibility of subject-doubling. Moreover, I assume that this is an ongoing change in Udmurt syntax. That is why the person agreement suffixes on the converbs show mixed properties: they seem very pronominal in terms of referentiality (for all speakers), but their pronominal nature is weakened in some varieties of Udmurt in which subject-doubling is allowed. The change might have started because of the fact that Udmurt is generally a pro-drop language. The idea that the pro-drop properties of a language might enforce the reanalysis of the former pronominal agreement markers into grammatical agreement is not new. A recent theoretical proposal for this type of diachronic development has been made by Fuß (2005: 141), who argues that the reanalysis is supported by the fact that speakers have a null pronoun at their disposal. As he suggests, the original pronominal agreement markers are reanalyzed as grammatical agreement markers that agree with a null pronoun. After the reanalysis, the former pronominal agreement marker can be occasionally doubled by an overt pronoun for the sake of emphasis, for instance.

At this point, a remark must be made regarding the internal syntax I assume for the converbial clauses in question. In the beginning of section 3, it was noted that pronominal agreement markers are sometimes said to be the arguments of the verb, hence there is no covert subject. However, others have claimed that the pronominal agreement affixes agree just as grammatical agreement markers do, the only difference being that the subject is covert. The choice between the two analyses is often language-specific and also sometimes theoretically motivated. For instance, Toivonen (2000) has proposed an LFG account of the Finnish possessive suffixes. In her account, no covert pronominals are used (this is a general principle in LFG). Moreover, she assumes that the possessive suffixes can have two lexical entries: one as an agreement marker (when accompanied by an overt pronominal possessor) and one as an incorporated pronoun (when it occurs on its own). Toivonen's (2000) analysis focuses mainly on general possessor constructions, while non-finite clauses are only touched upon as a further extension of her proposal. Finnish has a rich inventory of non-finite clauses and it is claimed that most of them: "can host both the agreement suffixes and the pronominal suffixes, parallel to the suffixes on nominals [...]" (Toivonen 2000: 605). However, she mentions that in the case of the so-called participles, only the pronominal suffixes can be used.

However, others have rejected this proposal (Huhmarniemi & Brattico 2015). They propose a unified analysis according to which the possessive suffixes are always agreement markers. Moreover, they assume a *pro* possessor and claim that the conditions for pro-drop are similar in finite clauses and in possessive constructions. Their work also mainly deals with the possessor suffixes in possessive constructions. Both analyses speculate about the behaviour of the possessive suffixes in non-finite clauses. Toivonen (2000: 605) stipulates that participles: "can host pronominal suffixes, but not agreement suffixes." Huhmarniemi & Brattico (2015), on the other hand, propose that the non-finite clauses are control structures with a PRO subject.

Here I do not aim to evaluate these two proposals in the case of Finnish. Moreover, the Finnish non-finite clauses discussed in these two papers are not comparable to the Udmurt converbial clauses since they are complement clauses. It is a well-known fact that complement and adjunct non-finite clauses differ considerably from one another, for instance with respect to agreement marking and control possibilities. Hence, I will not make parallels with the relevant Finnish non-finite clauses.

As far as the theoretical background is concerned, I will take the pro-drop approach. I assume that the converbial clauses have a null subject with which the person agreement markers agree. However, for the time being, I would not like to speculate about the exact category of the null subject (*pro* or PRO) since this question deserves further investigation. A similar analysis has been proposed by Tóth (2011) in the case of Hungarian inflected infinitives. However, in Udmurt, the null subject must be an argument, as we have seen in 3.1. Even quasi-argumental subjects in the case of weather predicates are impossible.

This approach has several advantages. First of all, it assumes that examples like (2), (3), and (4) have a unified structure, the only difference being that the subject in (3) is overt, while in (2) and (4), it is covert. If no null subject is assumed in (4), it raises the question of how these clauses can have an overt one (as in (3)). The null subject is syntactically active, for instance, with regard to binding. Moreover, if there was no null subject in examples like (2), it would be hard to explain why speakers of variety B would allow subject-doubling. On the contrary, if a null subject is assumed, one can state that these speakers simply reinforce the null pronoun for the sake of emphasis. It is a well-known fact that null pronouns cannot be used to express focus, for instance. It is also plausible that the person agreement markers on the converbs—even if we regard them as incorporated pronouns—cannot be used for emphasis since they are not phonologically independent words.

The proposed morphosyntactic change is also supported by the typological evidence. Siewierska (2004: 126) distinguishes three types of agreement markers: pronominal, ambiguous, and syntactic. Ambiguous markers are the instances where the markers can be, but not necessarily, doubled by an overt DP (as in pro-drop languages). Cross-linguistically it is very common for grammatical agreement to originate from pronominal agreement markers (Siewierska 2004: 262). Hence, I propose that in Udmurt, the former pronominal suffixes are being reanalyzed as instances of pro-drop. Since in Udmurt the finite clauses can have pro-drop, the speakers analogically reanalyze the non-finite clauses as having pro-drop, too. I argue that the morphosyntactic change has gone even further for the speakers of variety B (the ones who allow subject-doubling). For these speakers, the pronominal nature of the suffixes is weakened to the extent that an overt subject DP can co-occur with the person agreement suffix.

A further piece of evidence in favour of this scenario is that the change does not affect only the converbs, but also other agreement targets on which the same suffixes are used, for instance postpositions. In example (29), the 1SG agreement suffix is doubled by an overt 1SG pronoun.

(29) (Edygarova 2010: 142)
Vot *tare* [*mon* *śar̯iśt-i̯m*] *ki̯čak* *tod-ko-di̯.*
well then 1SG about-1SG in.detail know-PRS-2PL
'Well, then you know all about me.'

Edygarova (2010: 142) claims that such doubling is not characteristic of standard Udmurt and considers examples like (29) to be dialectal. It should be pointed out that the examples she presents are all with pronouns, and not with lexical DPs. Nevertheless, her examples support my claim that the person agreement suffixes used on the converbs, postpositions, and adverbs are being reanalyzed as grammatical agreement.

It should be also considered whether this change in the grammar of Udmurt is influenced by Russian. It is a well-known fact that Russian has influenced/has been influencing Udmurt in very many respects (lexicon, word order). Russian is a non-pro-drop language and we might suppose that Udmurt might start or might have already started using overt pronouns more often due to Russian influence. It can be also suggested that Russian influence is one of the reasons for the reanalysis of the person agreement on the Udmurt converbs as well. In my view, the Russian influence on Udmurt syntax can never be excluded, but in the case of the converbs, the change can be explained as a language-internal one, and it also follows a typologically observed path.

One more peculiarity remains to be explained. Recall that the verbal agreement, the possessive suffixes and the person agreement on the converbs are very similar and probably are of the same origin. Interestingly enough, neither the verbal agreement on the finite verbs nor the possessive suffixes seem to be pronominal in nature. In section 3.1., it was shown that the agreement on the converbs is highly referential, whereas the agreement on the finite verbs is not. Moreover, subject-doubling is perfectly possible in finite clauses for all of the speakers, but not in non-finite ones. It might tentatively be proposed that originally, the verbal agreement and possessive suffixes were pronominal, too. Moreover, one could speculate that the agreement in the non-finite clauses has remained pronominal because non-finite clauses, or in other words embedded clauses, are generally less susceptible to syntactic changes than finite main clauses. More specifically, the grammaticalization of agreement markers has to do with word order and information structure. As Givón (1976) suggests, the rise of agreement is actually a reanalysis of a topical DP constituent as a subject, and at the same time, the reanalysis of the original pronouns/pronominal agreement markers as grammatical agreement markers. In this vein, I would suggest that the reanalysis could take place only in the finite clauses since only the finite TPs can be merged with information structural projections. Moreover, Udmurt finite clauses have free word order, which makes the reanalysis possible. Udmurt non-finite clauses, however, could not undergo this change because they have rigid SOV word order and probably cannot be merged with information structure projections at all.

4 Conclusion

In this paper, I have discussed the person agreement used on Udmurt converbs formed with the suffixes *-ku* and *-toź*. I have argued that there is an ongoing change in the grammar of Udmurt with respect to these two converbs. This change involves the reanalysis of the former pronominal agreement on the converbs into grammatical agreement with pro-drop. I have shown that person agreement on the converbs has mixed properties. With respect to referentiality, it is very pronominal in nature. This means that the person agreement markers can agree with referential null subjects only (even quasi-argumental subjects of weather predicates are excluded). On the other hand, many speakers allow subject-doubling of the originally pronominal suffixes.

This suggests that for those speakers, the person agreement on the converbs is not used strictly as pronominal agreement, but rather has become grammatical agreement with pro-drop. I have demonstrated the variation among the speakers of Udmurt with regard to subject-doubling, as well as the variation of the subject-doubling possibilities. This extensive variation indicates that we are dealing with an ongoing process. This change is motivated by the fact that Udmurt is a pro-drop language and the Udmurt non-finite clauses are being reanalyzed analogically to their finite counterparts as having a dropped subject. The change also fits nicely with the typological observation on how agreement arises. In this way, this paper sheds light on the some unexplored issues of Udmurt morphosyntax and contributes to our general knowledge on the Udmurt language and its varieties.

Abbreviations

ADD	additive
ACC	accusative
ALL	allative
AUX	auxiliary
DAT	dative
DET	determiner
EXCL	exclusive
EVID	evidential
FREQ	frequentative
FUT	future
GEN	genitive
ILL	illative
INES	inessive
INF	infinitive
INS	instrumental
INTR	intransitive
NEG	negation
NF	non-finite
PL	plural
PN	proper noun
PRES	present
PST	past
SG	singular
TERM	terminative

References

Abney, Steven P. 1987. *The English noun phrase in its sentential aspect*. Cambridge, MA: Massachusetts Institute of Technology PhD dissertation.

Alexiadou, Artemis. 2001. *Functional structure in nominals: nominalization, and ergativity*. (Linguistik Aktuell/Linguistics Today, vol. 42). Amsterdam: John Benjamins. doi: http://dx.doi.org/10.1075/la.42.

Anderson, Stephen R. 2005. *Aspects of the theory of clitics*. Oxford: Oxford University Press. doi: http://dx.doi.org/10.3366/E1750124509000348.

Arkhangelskiy, Timofey A. 2014. Clitics in the Beserman dialect Of Udmurt. *Higher School of Economics Research Paper No. WP BRP* 10. https://www.hse.ru/pubs/share/direct/document/178195497 (2 June, 2016.)

Bartens, Raija. 2000. *Permiläisten kielten rakenne ja kehitys*. Helsinki: Suomalais-Ugrilainen Seura.

Bíró, Bernadett. 2011. Cselekvésnevek a manysiban—az északi manysi cselekvésnevek valenciája. *Folia Uralica Debreceniensia* 18. 3–33.

Borsley, Robert & Jaklin Kornfilt. 2000. Mixed extended projections. In Robert Borsley (ed.), *The nature and function of syntactic categories*, 101–131. Leiden: Brill. doi: http://dx.doi.org/10.1163/9781849500098.

Bresnan, Joan & Sam A. Mchombo. 1987. Topic, pronoun and agreement in Chichewâ. *Language* 63(4). 741–782. doi: http://dx.doi.org/10.2307/415717.

Chafe, Wallace. 1994. *Discourse, consciousness, and time: The flow and displacement of conscious experience in speaking and writing*. Chicago: The University of Chicago Press.

Comrie, Bernard & Sandra Thompson. 1985. Lexical nominalizations. In Timothy Shopen (ed.), *Language typology and syntactic description Volume III, Grammatical categories and the lexicon*, 349–398. Cambridge, UK: Cambridge University Press.

Corbett, Greville. 2006. *Agreement*. Cambridge: Cambridge University Press.

Csúcs, Sándor. 1988. Die wotjakische Sprache. In Denis Sinor (ed.), *The Uralic Languages: description, history, and foreign influences* (Handbuch Der Orientalistik Achte Abteilung Handbook of Uralic Studies, vol. 1), 131–146. Leiden & New York: Brill.

Dékány, Éva. 2014. A nem véges alárendelés (az igenevek) története. In Katalin É. Kiss (ed.), *Ómagyar generatív szintaxis*, 179–239. Budapest: Akadémiai Kiadó.

Edygarova, Svetlana. 2010. *Kategoria posessivnosti v udmurtskom yazyke*. Tartu: University of Tartu PhD dissertation.

Evans, Nicholas. 1999. Why argument affixes in polysynthetic languages are not pronouns: evidence from Bininj Gun-wok. *STUF-Language Typology and Universals* 52(3–4). 255–281. doi: http://dx.doi.org/10.1524/stuf.1999.52.34.255.

Fuß, Eric. 2005. *The Rise of Agreement. A formal approach to the syntax and grammaticalization of verbal inflection* (Linguistik Aktuell/Linguistics Today, 81). Amsterdam & Philadelphia: John Benjamins. doi: http://dx.doi.org/10.1075/la.81.

Georgieva, Ekaterina. 2012. *Igeneves szerkezetek az udmurt nyelvben*. Szeged: University of Szeged MA thesis.

Givón, Talmy. 1976. Topic, pronoun, and grammatical agreement. In Charles N. Li (ed.), *Subject and topic*, 149–188. New York: Academic Press.

Grimshaw, Jane B. 1990. *Argument Structure* (Linguistic Inquiry Monographs 18). Cambridge, MA: Massachusetts Institute of Technology.

Haspelmath, Martin. 1995. The converb as a cross-linguistically valid category. In Martin Haspelmath & Ekkehard König (eds.), *Converbs in cross-linguistic perspective. Structure and meaning of adverbial verb forms—adverbial participles, gerunds.* (Empirical Approaches to Language Typology 13), 1–55. Berlin & New York: Mouton de Gruyter.

Haspelmath, Martin. 1996. Word-class-changing inflection and morphological theory. In Geert Booij & Jaap van Marle (eds.), *Yearbook of morphology 1995*, 43–66. Dordrecht: Kluwer.

Huhmarniemi, Saara & Paoli Brattico. 2015. The Finnish possessive suffix. *Finno-Ugric Languages and Linguistics* VI(1–2). 2–41. http://full.btk.ppke.hu/index.php/FULL/article/view/31 (2 June, 2016.)

Jelinek, Eloise. 1984. Empty categories, case, and configurationality. *Natural Language & Linguistic Theory* 2(1). 39–76.

Karpova, Ludmila L. 2005. *Srednečepeckij dialect udmurtskogo yazyka: obrazcy reči.* Izhevsk: Udmurtskij institut istorii, yazyka i literatury UrO, RAN.

Kelmakov, Valentin K. 1998. *Kratkij kurs udmurtskoj dialektologij.* Izhevsk: Izdateľstvo Udmurtskogo universiteta.

Koptjevskaja-Tamm, Maria. 1993. *Nominalizations.* London: Routledge.

Koptjevskaja-Tamm, Maria. 2003. Action nominal constructions in the languages of Europe. In Frans Plank (ed.), *Noun phrase structure in the languages of Europe*, 723–761. Berlin & New York: Mouton de Gruyter.

Kornfilt, Jaklin. 2003. Subject Case in Turkish nominalized clauses. In Uwe Junghanns & Luka Szucsich (eds.), *Syntactic structures and morphological information* (Interface Explorations 7), 129–216. New York: Mouton De Gruyter.

Laczkó, Tibor. 2000. Az ige argumentumszerkezetét megőrző főnévképzés. In Ferenc Kiefer (ed.), *Strukturális magyar nyelvtan 3. Morfológia*, 293–407. Budapest: Akadémiai Kiadó.

McCloskey, James & Kenneth Hale. 1984. On the syntax of person-number inflection in Modern Irish. *Natural Language & Linguistic Theory* 1(4). 487–533. doi: http://dx.doi.org/10.1007/BF00417057.

Mithun, Marianne. 2003. Pronouns and agreement: The information status of pronominal affixes. *Transactions of the Philological Society* 101(2). 235–278. doi: http://dx.doi.org/10.1111/1467-968X.00119.

Nikolaeva, Irina. 1999. *Ostyak* (Languages of the World/Materials 305). München & Newcastle: LINCOM EUROPA.

Ótott-Kovács, Eszter. 2015. *The syntax of non-finite clauses in Kazakh*. Szeged: University of Szeged PhD dissertation.

Perevoščikov, Petr N. (ed.) 1962. *Grammatika sovremennogo udmurtskogo yazyka. I. Fonetika i morfologija*. Izhevsk: Udmurtskoe Knižnoje Izdatel'stvo.

Perevoščikov, Petr N. 1959. *Deepričastiya i deepričastnye konstrukcii v udmurtskom yazyke*. Izhevsk: Udmurtskoe Knižnoje Izdatel'stvo.

Raun, Alo. 1988. Proto-Uralic comparative historical morphosyntax. In Denis Sinor (ed.), *The Uralic languages: description, history and foreign influences* (Handbuch Der Orientalistik Achte Abteilung Handbook of Uralic Studies, vol. 1), 555–571. Leiden & New York: Brill.

Riese, Timothy. 1998. Permian. In Daniel Abondolo (ed.), *The Uralic languages* (Routledge Language Family Descriptions), 249–275. London & New York: Routledge.

Sahan, Fatma. 2002. *Nominal clauses in Kazan Tatar*. Madison, WI: University of Wisconsin-Madison PhD dissertation.

Sárik, Pál. 1998. A határozói igenevek néhány problémája. *Magyar Nyelv* 94. 423–436.

Serdobolskaja, Natalia M., A.A. Il'evskaya, S.A. Minor, P.S. Miteva, A.V. Fajnvejc & N.S. Matveeva. 2012. Konstrukcii s sentencial'nymi aktantami v finno-ugorskix jazykax. In Ariadna I. Kuznecova (ed.), *Finno-ugorskie jazyki: fragmenty grammatičeskogo opisanija. Formal'nyj i funkcional'nyj podxody*, 382–476. Moskva: Rukopisnye pamyatniki Drevnej Rusi.

Siewierska, Anna. 1999. From anaphoric pronoun to grammatical agreement marker: Why objects don't make it. *Folia Linguistica* 33(1–2). 225–252.

Siewierska, Anna. 2004. *Person*. Cambridge: Cambridge University Press. doi: http://dx.doi.org/10.1017/CBO9780511812729.

Teplyashina, Tamara I. 1970.*Yazyk besermyan*. Moscow: Nauka.

Toivonen, Ida. 2000. The morphosyntax of Finnish possessives. *Natural Language & Linguistic Theory* 18(3). 579–609.

Tóth, Ildikó. 2000. *Inflected Infinitives in Hungarian*. Tilburg: Tilburg University PhD dissertation.

Tóth, Ildikó. 2011. Non-referential readings of null subjects in Hungarian. In Tibor Laczkó & Catherine O. Ringen (eds.), *Approaches to Hungarian: Volume 12: Papers from the 2009 Debrecen Conference*, 209–238. Amsterdam, Philadelphia: John Benjamins. doi: http://dx.doi.org/10.1075/atoh.12.08tot.

Winkler, Eberhard. 2001. *Udmurt* (Languages of the World/Materials 212). München & Newcastle: LINCOM EUROPA.

Ylikoski, Jussi. 2003. Defining non-finites: action nominals, converbs and infinitives. *SKY Journal of linguistics* 16(1). 185–237.

Zwicky, Arnold M. & Geoffrey K. Pullum. 1983. Cliticization vs. Inflection: English N'T. *Language* 59(3). 502–513.

CHAPTER 5

Paradigm Leveling: The Decay of Consonant Alternations in Russian*

Varvara Magomedova and Natalia Slioussar

In this paper, we look at a paradigm leveling process currently taking place in Russian that affects historical consonant alternations (morphophonemic alternations that arose as a result of historical sound changes in Slavic and Russian specifically). In Standard Russian, these alternations are present in some verb forms (*ljubit'* 'to love'—*ljublju* 'I love'), in comparatives (*suxoj* 'dry'—*suše* 'drier, more dryly'), in deverbal nouns, and in some other grammatical categories. However, many non-standard forms in Russian lack alternations or have 'incorrect' alternations unattested in the standard language. Unfortunately, Russian corpora contain almost no such non-standard forms, and the best source of such data is the Internet. However, estimating relative frequencies of different forms found on the Internet is a challenge because the counts provided by search engines are extremely unreliable. We developed various strategies and a program to circumvent this problem and applied the technique to our study of alternations primarily in comparatives and to some extent also in verb forms.

1 Introduction

The Internet gives us access to a massive volume of unedited texts and thus allows us to investigate the dynamics of language change on a new level. In this paper, we use this opportunity to analyze the decay of historic consonant alternations found at the end of some stems in Russian (here and further, we use transliteration rather than transcription and make additional comments wherever necessary):

* This study was partially supported by the Basic Research Program of the National Research University Higher School of Economics. We are grateful to several linguists for their valuable comments, in particular, to Christina Bethin, Ellen Broselow, and Michael Becker.

 | DOI 10.1163/9789004342934_007

(1) *g / k / x* before *e, i* > *ž / č / š* (Slavic first velar palatalization) or palatalized *z / c /* palatalized *s* (Slavic second palatalization)
g / k / x + *j* > *ž / č / š*
d / t + *j* > *ž / č* or *žd / šč*
z / s + *j* > *ž / š*
b / p / v / m + *j* > *bl / pl / vl / ml* (with palatalized *l*)
st / sk / kt / gt + *j* > *č* or *šč*

In Old Russian, these alternations were present in various inflectional paradigms and in derivation. In the standard variety of Modern Russian, the number of these alternations has decreased significantly, but the remaining ones are stable: they are present in several verb classes, in comparatives and before certain derivational suffixes (e.g. *brosit'* 'to throw'—*brošu* 'throw$_{\text{FUT.1SG}}$', *molodoj* 'young'—*molože* 'younger', *dvigat'* 'to move'—*dviženie* 'movement'). However, in many non-standard forms found on the Internet these alternations are either absent or distorted.

We have two goals in this paper: to identify various factors that influence the decay of consonant alternations and to discuss how we analyzed the Internet data, presenting the program we created to facilitate work with Internet search results.

In our view, investigating this process is especially interesting because relatively little is known about the dynamics of language change in the era of mass literacy. Such processes are slowed down dramatically but are not completely frozen, and they manifest themselves in various domains where the pressure of the literary norm is lower. Another matter of general interest is to develop tools and techniques to work with Internet data. Linguists have never had access to such a large volume of unnormalized texts before and it can serve as a rich new source of material for language change studies. We only scratch the surface of this problem here, more work remains to be done.

We start our discussion with alternations in verbs and then turn to comparatives. The decay of alternations in verb forms will be presented in less detail because these facts have been analyzed in our previous work (Slioussar and Kholodilova, 2013). In the future, we are also going to analyze the fate of alternations before certain derivational suffixes. This study was partially supported by the Basic Research Program of the National Research University Higher School of Economics.

2 Alternations in Verb Forms

Several verb classes have these types of alternations in Standard Russian. Below we will classify them according to the One Stem System proposed by

Jakobson (1948) and his followers.[1] Classes I and E have alternations in the 1sg present/future form and in the passive past participle: e.g. *brosit'* 'to throw'—*brošu* 'throw$_{\text{FUT.1SG}}$,' *brosiš* 'throw$_{\text{FUT.2SG}}$' etc.—*brošennyj* 'thrown$_{\text{M.NOM.SG}}$.' Class A has alternations in all present/future forms and in the active present participle: e.g. *pisat'* 'to write'—*pišu* 'write$_{\text{PRS.1SG}}$,' *pišeš* 'write$_{\text{PRS.2SG}}$' etc.—*pišuščij* 'writing.' The pattern in the class G-K is complicated: many forms have alternations.

The situation with alternations in verb forms has already been discussed in our earlier work (Slioussar and Kholodilova, 2013), so we only present the main findings here. The I-stem class is the only productive one having alternations, and many new slang verbs have recently appeared as I-stem verbs, for example, *frendit'* 'to include in the friend list,' *kommentit'* 'to post comments' etc. This is the class with the largest number of non-standard forms, and Slioussar and Kholodilova (2013) demonstrated that several factors influence the distribution of different forms with and without mutations. First, less frequent verbs are more likely to lack alternations. Second, verbs lack alternation more often if they do not belong to the standard language. Third, the final consonant of the stem matters: stems ending in obstruent clusters lose alternations the most often, while stems ending in labials hardly have forms lacking alternations.

In the G-K class (*peč* 'to bake'—*peku* 'bake$_{\text{PRS.1SG}}$,' *pečëš* 'bake$_{\text{PRS.2SG}}$') there are two kinds of non-standard forms: those lacking alternations and those having alternations, although we expect there should be none. Notably, alternation loss occurs much more often than alternation overuse although there are more forms with alternations in the present/future tense paradigm. This fact contradicts McCarthy's (2005) theory as he suggests that paradigm leveling should go in the direction of overapplication of alternations. Our data can be better accommodated by other models (e.g. Albright, 2010).

In the A-stem class (e.g. *pisat'* 'to write'—*pišu* 'write$_{\text{PRS.1SG}}$'), there is no alternation decay per se, but some verbs lose alternations because they migrate to the productive AJ class where there are no alternations (e.g. *maxat'* 'to wave'—*mašu* 'wave$_{\text{PRS.1SG}}$,' *maxaju* 'wave$_{\text{PRS.1SG}}$'). This process is thoroughly analyzed in several works (e.g. Nesset and Kuznetsova, 2010; Nesset and Janda, 2011).

Notably, some non-standard forms have alternations unattested in standard Russian, e.g. not *d // ž*, but *d // č*, *d /// š*, *d // dž* etc. In other words, speakers still know that there should be an alternation, but already have doubts about which alternation to use. This may have important implications for various models of the mental lexicon.

1 Several approaches to Russian verb classification exist, but we will not discuss them here because this is not important for our paper and all classes will be illustrated by examples.

3 Alternations in Comparatives

3.1 *Comparative Formation: Outlining the Data Set*

Comparatives can be derived from some adjectives and adverbs. As our research focuses on the decay of morphophonemic alternations, and we study comparatives only from this perspective, it is not important for us whether a comparative comes from an adverb or an adjective, especially since most adverbs that have comparative forms are derived from adjectives. Therefore, when we prepared our data set, we focused mostly on adjectives.

There are four types of comparatives in modern Russian:

- uninflected synthetic forms (e.g. *moložе* from *molodoj* 'young,' *bol'še* from *bol'šoj* 'big');
- uninflected synthetic forms with a *po-* prefix, which introduces the meaning 'slightly, somewhat' (e.g. *pomolože, pobol'še*);
- analytic forms (e.g. *bolee molodoj, bolee bol'šoj*);
- declinable synthetic forms, which can be formed only from a couple of adjectives (e.g. *bol'šij* from *bol'šoj* 'big,' *men'šij* from *malen'kij* 'small').

Many adjectives have comparative forms of three first types. Their usage depends on a number of syntactic and stylistic factors, and they may have different nuances of meaning. In particular, synthetic comparatives cannot be used attributively because they are uninflected, but they are preferred as predicates. This question is explored in great detail by Knjazev (2007). Obviously, only synthetic comparatives may have consonant alternations, so these are the focus of our paper.

Forming synthetic comparatives from some adjectives is problematic or impossible in Standard Russian. This may be due to semantic or morphological reasons. In the former case, neither synthetic nor analytic forms are used because the semantics of the adjectives are incompatible with comparative formation.[2] In the latter case (examples of such adjectives are given in Table 5.1 below), analytic forms should be used. However, native speakers also tend to generate non-standard synthetic forms, some of which are of special interest for our study.

2 Notably, many adjectives have a primary meaning incompatible with comparative formation and a secondary meaning that allows for it. For example, when *stekljannyj* 'made out of glass' is used metaphorically in the expression *stekljannyj vzgljad* 'glassy eyes,' literally 'glassy gaze,' we can say about somebody: *Ego vzgljad stal ečše stekljannee* 'His eyes became even glassier.'

Synthetic comparatives can be formed with three suffixes in Russian: *-e* (e.g. *molože* from *molodoj* 'young'), *-ee/ej* (e.g. *veselee* or *veselej* from *vesëlyj* 'merry')[3] and *-še* (e.g. *ton'še* from *tonkij* 'thin'). The suffix *-še* is very rare, while *-ee/ej* is the most frequent and most recent diachronically. Only comparatives with the *-e* suffix have consonant alternations.

The suffix *-e* can be added to primary adjectives that do not have any other suffixes (i.e. consist of the root and an ending) and to the adjectives with a *-k-/-ok-* suffix, which is dropped in some comparative forms (e.g. *niže* from *nizkij* 'low') and remains intact in the others (e.g. *žutče* from *žutkij* 'terrible'). With a couple of exceptions like *deševle* from *dešëvyj* 'cheap,' the *-e* suffix is used with two groups of stems: ending in velar consonants (*-g, -k, -x*) and in *-t* or *-d*. As Table 5.1 shows, these two groups are quite different, and we compare them in our study.

In the first group, adjectives either have an alternation in the comparative form or have no synthetic comparative at all. Adjectives from the second group mostly have comparatives with the *-ee/ej* suffix and no alternation; only ten adjectives use the *-e* suffix. The counts in Table 5.1 are based on the *Grammatical Dictionary of Russian*[4] (Zaliznjak 1977) (Zaliznjak outlines some generalizations and exceptions in his extensive introduction to the dictionary, while all other information including the counts was extracted from an electronic database based on the dictionary).

We chose two groups of comparatives for our study: the ones that have a normative synthetic form and the ones that do not. As Table 5.1 shows, there are 123 adjectives that have standard comparatives with alternations in the *Grammatical Dictionary of the Russian Language* (Zaliznjak 1977). Non-standard forms are relatively frequent in some of these cases and very rare in others.[5] So we narrowed down the group of adjectives in the following way. For every adjective, we searched on the Internet for the comparative form with the suffix *-ee* without consonant alternations. If in the first thousand hits (or fewer, if the search returned fewer) there were more than 4% comparatives and no typos or other irrelevant data, we selected the adjective for further study. This way, 23 out of 113 adjectives with velar stems and nine out of ten adjectives with

3 In principle, *-ej* is a colloquial variant of *-ee*, but it can be used in poetry when the meter requires it.

4 The dictionary contains around 100 000 words and is often used as the most extensive source of lemmas in Russian.

5 As Slioussar and Kholodilova (2013) demonstrated in their study discussed in section 2, the situation with the verbs is similar: non-standard forms from verbs that belong to literary Russian also tend to be very rare.

TABLE 5.1 *Comparatives from different types of adjectives according to Zaliznjak (1977)*

	Stems ending in *-g, -k, -x*	Stems ending in *-t, -d*
Comparatives with *-e*	113 adjectives forming comparatives with alternations.	Ten adjectives forming comparatives with alternations.
Comparatives with *-ee/ej*	Only one adjective *merzkij* 'nasty' has a form *merzee* (*merzče* is an alternative standard comparative). No forms end in *-gee, -kee* or *-xee.*	More than 1000 adjectives, simple (e.g. *žëltyj* 'yellow') and compound (e.g. *tolstomordyj* 'fat-faced').
No synthetic comparatives	– More than 3500 adjectives with suffixes *-sk-, -ck–* (relational) and *-en'k-, -on'k–* (diminutive); – About 140 compound adjectives (e.g. *dlinnonogij* 'long-legged'); – About 40 rare adjectives that have a short form only (*slabëxonek* 'weak'); – *Blagoj* 'good', *velikij* 'great', *kačkij* 'pitching', *nagoj* 'nude', *pegij* 'piebald'.	– *Gordyj* 'proud', *zanjatyj* 'busy', *pripodnjatyj* 'elevated', *prinjatyj* 'accepted', *chuždyj* 'alien'; – Compound adjectives denoting color (e.g. *bledno-žëltyj* 'pale yellow').

dental stems[6] were selected for the first group. They are listed in Tables 5.2 and 5.3 below.

For the second group we selected a number of adjectives that have no standard synthetic comparatives: 13 compound adjectives, six adjectives with *-sk-* or *-ck-* suffixes and and *velikij* 'great'—the only suffixless adjective without a standard synthetic comparative in the velar group that has enough non-standard forms. These adjectives are listed in Tables 5.4 and 5.5 below. Notably, in some compound adjectives, the second part of the word can be used as an independent adjective (e.g. *zorkij* 'sharp-sighted' for *dal'nozorkij* 'long-sighted,'

6 The only adjective with the dental stem we excluded was discarded before the preliminary analysis. *Xudoj* meaning 'slim' forms the comparative *xudee*, while *xudoj* in the almost obsolete meaning 'bad' forms the comparative *xuže*, which is now used as a suppletive form of *ploxoj* 'bad.' Thus, *xuže* and *xudee* cannot be licitly compared.

plavkij 'meltable' for *legkoplavkij* 'easily meltable'). However, these short adjectives are relatively infrequent—none of them was included in the *Frequency Dictionary of the Modern Russian Language* (Lyashevskaya and Sharoff 2009). In the other compound adjectives, the second part of the word cannot be used independently (e.g. *dlinnonogij* 'long-legged,' *dlinnorukij* 'long-armed'). However, the relevant stems can be found in numerous highly frequent words, many of them with consonant alternations. For example, according to the *Frequency Dictionary of the Modern Russian Language*, *ručka* 'small arm, small hand, handle' has the frequency of 57,6 per million, *nožka* 'small leg, furniture leg'—31,9 per million. We wanted to see how these factors influence the use of alternations.

3.2 *Results and Discussion*

As we mentioned in the introduction, we used the Internet to look for non-standard forms. The general principles we relied on (how to estimate frequencies when searching on the Internet, etc.) and the program we developed to facilitate our task are presented in section 4. Here we discuss only what we searched for and what we found.

At the first stage, we had to identify the pool of possible forms, so that their relative frequency could be estimated at the second stage. In particular, it should be noted that comparatives are affected not only by the tendency to eliminate a consonant alternation that reveals itself in non-standard forms. Another tendency is also at work: an older *-e* suffix is being replaced by *-ee/ej*. *Prima facie* these two processes should go hand in hand because *-ee/ej* never triggers consonant alternations. However, our preliminary Internet searches returned various examples with the *-ee/ej* suffix in which the alternation is retained so we included them in the pool. At the same time, we could find only a couple of examples with the suffix *-e*, but without an alternation like *krute* from *krutoj* 'steep, cool.' Given that they are so few, we cannot be sure that they are not typos (and what was meant was either the correct form *kruče* or a non-standard form *krutee/krutej*, which can be found at least in tens of thousands). Therefore, these forms were not included in our systematic searches at the second stage.

We also did not include forms with alternations unattested in Standard Russian, like *krušče* from *krutoj*. There are very few of them, and we could find only examples with *t // šč* (this phenomenon is much more widespread and diverse in verb forms, which were discussed in section 2). If an adjective has the suffix *-k-/-ok-*, it may be retained or dropped in non-standard forms. Given all these possibilities and the fact that synthetic comparatives can be used with the *po-* prefix, we looked for 10 different comparative forms for every suffixless adjective we studied and the 14 forms of every adjective with the suffix *-k-/-ok-*, *-sk-*, or *-ck-*. This is exemplified in (2) for the adjective *nizkij* 'low.'

(2) *niže* (correct form, the *-k-/-ok-* suffix is dropped), *poniže* (correct form with the *po-* prefix), *nizkee, nizkej, ponizkee, ponizkej* (non-standard forms with the *-ee/ej* suffix, without alternations, with or without the *po-* prefix, the *-k-/-ok-* suffix is retained), *nizee, nizej, ponizee, ponizej* (non-standard forms with the *-ee/ej* suffix, without alternations, with or without the *po-* prefix, the *-k-/-ok-* suffix is dropped), *nižee, nižej, ponižee, ponižej* (non-standard forms with the *-ee/ej* suffix, with alternations, with or without the *po-* prefix, the *-k-/-ok-* suffix behaves as in the correct form).

We also found several forms with alternations and with the suffix *-e* in which the suffix *-k-/-ok-* is incorrectly dropped or retained, e.g. *nizče* from *nizkij* 'low' instead of *niže, skol'že* from *skol'zkij* 'slippery' instead of *skol'zče, derže* from *derzkij* 'daring, impertinent' instead of *derzče*. In very few cases, the same is possible with the *-ee/ej* suffix, e.g. *xliplee* from *xlipkij* 'slimsy' instead of *xlipče*. Despite the fact that the *-še* suffix is extremely infrequent, there are also a number of non-standard comparatives with this suffix, e.g. *derzše* with a dropped suffix from *derzkij* 'daring, impertinent' instead of *derzče*. All these forms are very rare, so we do not examine them further in this study.

After we decided for which forms to search, we analyzed their distribution for every adjective we selected, limiting ourselves to the first one thousand results. The outcome is presented in Tables 5.2–5.5. The forms are grouped in the following way: forms with the suffix *-e* that have alternations (if an adjective has a standard synthetic comparative at all, this is the correct form); forms with the suffix *-ee* or *-ej* without alternations (for the adjectives that have suffixes *-k-/-ok-, -sk-* or *-ck-*, forms where these suffixes are and are not dropped are counted separately); forms with the suffix *-ee* or *-ej* that have alternations. Forms with the prefix *po-* and without it are counted together. The relative frequency of the forms with the *-ej* suffix are on average two times lower than of the *-ee* forms. Forms with the *po-* prefix are on average ten times less frequent than forms with no prefix. They are most frequent in the case of *tugoj* 'tight': here comparatives with *po-* are only three times less frequent than forms with no prefix. In Table 5.4, compounds are divided into groups 1 and 2 (depending on whether the second part can or cannot be used as an independent adjective). Information about lemma frequency (in items per million) is taken from the *Modern Russian Language Frequency Dictionary* (Lyashevskaya and Sharoff 2009). If a particular word was not included in the dictionary, the field is left blank.

The results for the adjectives that have standard synthetic comparatives show that frequency and the final consonant of the stem were not significant factors, unlike with the verbs discussed in section 2. Rather, there are several individual adjectives that have problems with alternations (especially *ubogij, uprugij, derzkij, odinokij,* and *skol'zkij*), while for the rest, correct forms are used in the absolute majority of cases. We explain this by the fact that in the case of verbs, we studied primarily a productive class, namely the i-stem class, where the process of alternation loss takes a more systematic turn, while the groups of adjectives we analyze here are not productive, so the tendency is limited to 'knocking out' individual adjectives.

Let us add that Tables 5.2 and 5.3 do not reflect the fact that especially in the case of highly frequent adjectives like *krutoj* 'cool, steep' or *molodoj* 'young,' even 0.1% may mean thousands of non-standard forms on the Internet. It should also be noted that the forms in which the suffix *-k-/-ok-* is dropped are

TABLE 5.2 *Sufixless adjectives that have standard synthetic comparatives*

Adjective	Translation	Stem	Frequency	Results after filtering	Suffix *-e*	Suffix *-ee/ej*	Suffix *-ee/ej* with alt.
častyj	frequent	dental	114,6	1308	100,0%	0,0%	0,0%
krutoj	steep	dental	43,4	1809	99,7%	0,2%	0,1%
bogatyj	rich	dental	85,0	1873	99,7%	0,3%	0,0%
čistyj	clean	dental	159,0	1574	99,6%	0,3%	0,1%
molodoj	young	dental	414,1	1006	99,5%	0,5%	0,0%
tverdyj	solid	dental	59,5	615	99,0%	1,0%	0,0%
tolstyj	thick	dental	84,4	1087	99,0%	0,5%	0,6%
gustoj	thick	dental	47,2	500	98,0%	1,6%	0,4%
prostoj	simple	dental	275,3	1622	97,1%	0,2%	2,7%
tugoj	tight	velar	11,9	492	100,0%	0,0%	0,0%
jarkij	bright	velar	2,2	1701	100,0%	0,0%	0,0%
strogij	strict	velar	62,8	693	99,9%	0,1%	0,0%
gluxoj	deaf	velar	40,7	766	99,7%	0,0%	0,3%
ploskij	flat	velar	29,1	686	99,3%	0,6%	0,2%
pologij	sloping	velar	5,0	294	98,6%	0,7%	0,7%
suxoj	dry	velar	83,0	475	95,2%	0,0%	4,8%
uprugij	resilient	velar	10,7	588	68,9%	27,0%	4,1%
ubogij	poky	velar	11,7	342	48,8%	32,8%	18,4%

TABLE 5.3 *Adjectives with the suffix* -k-/-ok- *that have standard synthetic comparatives*

Adjective	Translation	*-k-/-ok-* dropped in the correct form	Frequency	Results after filtering	Suffix *-e*	Suffix *-ee/ej*	Suffix *-ee/ej, -k-/-ok-* dropped	Suffix *-ee/ej* with alt.
vysokij	high	yes	483,3	1502	99,9%	0,0%	0,0%	0,1%
glubokij	deep	yes	137,3	1172	99,8%	0,0%	0,0%	0,2%
blizkij	close	yes	206,8	1505	98,2%	0,0%	1,7%	0,1%
nizkij	low	yes	160,7	1197	98,5%	0,0%	1,4%	0,1%
tjažkij	grievous	no	26,0	349	97,1%	1,7%	0,0%	1,2%
xlipkij	slimsy	no	2,4	705	96,8%	0,4%	1,3%	1,5%
stojkij	resistant	no	10,1	494	92,9%	0,2%	0,0%	6,9%
veskij	weighty	no	4,9	257	89,5%	0,0%	0,0%	10,5%
žestokij	cruel	no	37,5	581	84,5%	2,4%	0,0%	13,1%
žutkij	spooky	no	26,1	828	79,4%	1,4%	5,7%	13,2%
gromozdkij	cumbersome	no	7,1	274	75,9%	14,6%	7,3%	2,2%
derzkij	daring	no	8,9	387	60,7%	0,0%	37,7%	1,6%
odinokij	single	no	44,5	162	48,8%	29,0%	0,0%	22,2%
skol'zkij	slippery	no	11,8	325	13,9%	0,9%	81,8%	3,4%

relatively frequent among non-standard forms (e.g. *nizee* as opposed to *nizkee* from *nizkij* 'low' having the correct form *niže*). They allow the speaker to avoid alternations without using *-kee/kej, -gee/gej, -xee/xej* segments unattested in standard Russian.

The difference between the adjectives that have a standard synthetic comparative and the ones that do not is dramatic ($\chi 2$ = 53.3, $p < 0{,}001$). The difference between the two subgroups of compound adjectives in Table 5.4 is also highly significant ($\chi 2$ = 88.2, $p < 0{,}001$). Notably, adjectives from the second subgroup where the second part cannot be used as an independent adjective, like *dlinnonogij* or *lopouxij*, have problems with alternations in the majority of synthetic comparatives, although various words where the relevant stems have alternations are highly frequent (e.g. *ručka* 'small hand, handle,' *nožka* 'small leg, furniture leg,' *uško* 'small ear, eyelet'). Unlike compounds, adjectives in Table 5.5 do not show a particular pattern. Apparently, although speakers sometimes try to derive synthetic comparatives from them, they are relatively clueless about how to proceed.

TABLE 5.4 *Compound adjectives without a normative synthetic comparative*

Adjective	Translation	Subgroup	Frequency	Results after filtering	Suffix -*e*	Suffix -*ee/ej*	Suffix -*ee/ee* with alt.
ènergoemkij	power-consuming	1	1,1	132	89,4%	6,8%	3,8%
legkoplavkij	fusible	1	–	24	83,5%	4,0%	12,5%
žiznestojkij	enduring	1	0,4	16	81,3%	12,5%	6,3%
tugoplavkij	refractory	1	0,4	43	76,7%	16,3%	7,0%
trudoëmkij	laborious	1	4,7	287	74,6%	19,2%	6,3%
iznosostojkij	hardwearing	1	–	65	70,8%	15,4%	13,9%
morozostojkij	frost-resistant	1	0,6	22	54,6%	18,2%	27,3%
termostojkij	heat-resistant	1	–	21	42,9%	33,3%	23,8%
blizorukij	myopic	2	3,5	26	15,4%	73,1%	11,5%
dlinnonogij	leggy	2	3,3	114	2,6%	80,7%	16,7%
lopouxij	lop-eared	2	1,2	53	1,9%	84,9%	13,2%
bezrukij	armless	2	1,3	18	0,0%	88,9%	11,1%
dlinnorukij	long-armed	2	0,6	19	0,0%	73,7%	26,3%

TABLE 5.5 *Other adjectives without a normative synthetic comparative*

Adjective	Translation	Frequency	Results after filtering	Suffix -*e*	Suffix -*ee/ej*	Suffix -*ee/ej*, -*sk*- or -*ck*- dropped	Suffix -*ee/ee* with alt.
velikij	great	276,3	87	6,9%	20,7%	–	72,4%
ispanskij	Spanish	21,1	10	14,0%	43,0%	0,0%	43,0%
russkij	Russian	530,5	435	4,6%	53,3%	39,6%	2,5%
aziatskij	Asian	10,3	26	3,8%	11,6%	7,7%	76,9%
ital'janskij	Italian	40,9	14	0,0%	71,4%	14,3%	14,3%
nemeckij	German	123,6	27	0,0%	92,6%	0%	7,4%
francuzskij	French	101,5	244	43,0%	39,3%	15,2%	2,5%

4 Collecting Data from the Internet

We could not rely on normalized corpora in our study because they contain very few non-standard forms and we were specifically interested in such forms. So we searched for standard and non-standard forms on the Internet, trying to determine which non-standard forms are attested and to estimate relative frequencies of different forms. To do so, we developed several techniques and a program *Lingui-Pingui*. The most popular Russian search engine is *Yandex*, so we chose it for our study. We worked with the *Yandex* API (http://api.yandex.ru/).

Obviously, our dataset was not balanced, as it is done in linguistic corpora, so we cannot draw any definitive conclusions from pairwise comparisons between standard and non-standard forms. To give an example, we cannot say that the form *odinoče* from *odinokij* 'lonely' is twice as frequent as *odinokee* because this obviously depends on the genre of the texts, and this is something we cannot control when searching on the Internet. Nevertheless, it is valid to compare the prevalence of various non-standard forms for different adjectives to identify the factors that play a role in the process of language change, which is exactly what is needed in a study like ours.

Let us explain how relative frequencies can be estimated in such cases. Simply searching for different forms one after another is useless because the counts provided by search engines are extremely unreliable. We used the method initially suggested in Slioussar and Kholodilova (2014), putting all forms in one query, as in (3).

(3) *"niže" | "poniže" | "nizkee" | "nizkej" | "ponizkee" | "ponizkej" | "nizee" | "nizej" | "ponizee" | "ponizej" | "nižee" | "nižej" | "ponižee" | "ponižej"*

Then we analyzed the relative frequency of various forms in the first one thousand results (or less, if there were less). Originally this limitation was due to the search engine capacity, but we have resolved this problem using the algorithm described by Maria Kholodilova (2013: 20–21). Nevertheless, we continue to work with the first one thousand results when we need to see the general picture and download more results only in exceptional cases.[7]

Yandex was not developed as a tool for linguistic research, so it is often inconvenient to work with the search engine directly. It takes a lot of time to

7 The *Yandex* support team confirmed that the first thousand of found results is representative in terms of relative frequencies if we count them this way, as search results are initially mixed by relevance and do not depend on the order we put different forms in a query.

prepare queries, send them, collect data, and then to refine and analyze the data collected. We created a program called *Lingui-Pingui* to solve these problems.[8] The program works in three stages. First it forms queries, sends them and downloads search results in a convenient form. Then it filters the results (the filters are set by the user) and counts statistics. The last step is to sort the results and to recalculate the statistics.

At the first stage one can specify the list of morphs (prefixes, suffixes, and stems) and search for all possible combinations of these morphs on the Internet. All *Yandex* advanced search options (excluding strings with particular words from the results, etc.) are also available through the program. When the search is finished, *Lingui-Pingui* makes a folder *Raw_Data* for data files with a subfolder *Summary*. For each query string, two data files are created. The first one contains the query string and the results of the search (i.e. strings with the query word forms). The second file lies in the *Summary* folder and contains statistic data (how many strings were downloaded, in total and for each query form, how many duplicates there are, how many strings were left after additional filtering, how many strings contain more than one query form etc.).

Raw results inevitably contain irrelevant data (words with typos, etc.). Going through all the results manually is very time-consuming work, so sorting results can significantly ease this process. Probably, there are some words that tend to co-occur with the queries. Thus, having these words in a result string dramatically increases the chances that this string is relevant. Or, just the opposite, there may be words that never co-occur with the needed form. If these words are specified in two lists, *Lingui-Pingui* can sort search results into three groups and recalculate statistics for each of them. There can be other "markers" of comparative, for example genitive noun forms, but the identification and isolation of these is harder to put into the program and it does not occur often enough to warrant the effort. As the manual refining remains inevitable, there is a simple instrument to ease this work too. For more detailed information about the program see Magomedova (2013).

5 General Discussion and Conclusions

The situation with alternations in verb forms described by Slioussar and Kholodilova (2013) can be summarized as follows. If a verb has a standard form

8 Working with *Lingui-Pingui* may save a lot of time, but before starting to use it one has to contact *Yandex.XML* and ask them for a certain amount of queries per day. For more information see http://xml.yandex.ru/.

with alternations, people will use it in the absolute majority of cases without any problems. Problems arise with novel verbs that do not belong to Standard Russian (e.g. *frendit'* 'to include in one's friend list') and with verbs that have paradigm gaps.

The group of adjectives forming comparatives with alternations is not productive: no novel words appear in it. So we predictably find fewer problems here than in the case of verb forms. Nevertheless, it must be noted that some adjectives (like *ubogij* 'poky' and *uprugij* 'resilient') have up to 30% forms without alternations. So far, we could not find a significant correlation with lemma frequency or the last consonant of the stem.

The process of alternation loss can be seen in its full strength when speakers try to form synthetic comparatives from adjectives that are not supposed to have them in Standard Russian, for example, compounds. Therefore all their synthetic comparatives are non-standard. In the group of compound adjectives that we studied the most important factor was whether the second part of the word can be used as an independent adjective. If it cannot, as in the case of *dlinnonogij* 'leggy,' *dlinnorukij* 'long-armed,' *bezrukij* 'armless,' *blizorukij* 'myopic,' *lopouxij* 'lop-eared,' the majority of comparatives lack alternations. This is surprising because the relevant stems can be found in many highly frequent words, many of them with consonant alternations (e.g. *ručka* 'small hand, handle,' *nožka* 'small leg, furniture leg,' *uško* 'small ear, eyelet'). Thus, it seems to be crucial whether a particular form from a particular word is listed in the mental lexicon, not whether the model is available. This is similar to our results with the verbs: in the i-stem class, the model is productive, however, in non-standard verbs there is much more variation and therefore forms without alternations.

We observed not only the tendency to lose alternations, but also the process of suffix change from older *-e* to newer *-ee/ej*. One would expect these two processes to run in parallel (resulting in forms like *molodee* from *molodoj* 'young' instead of *molože*). However, our search results show that alternations are kept in a number of *-ee/ej* forms and these forms are even more frequent than forms lacking alternations. The opposite is impossible: there are no forms with the *-e* suffix lacking alternations. There are no innovative non-standard forms with this suffix. It is interesting to compare it with the suffix *-še* that is present only in a couple of normative forms and is still used in quite a few non-standard comparatives (e.g. *stran'še* from *strannyj* 'strange' instead of *strannee*, *derzše* from *derzkij* 'impudent' instead of *derzče*). Apparently, although both *-e* and *-še* suffixes are unproductive according to the standard criteria, the two have a different status in the mental lexicon.

References

Albright, Adam. 2010. Base-driven leveling in Yiddish verb paradigms. *Natural Language and Linguistic Theory* 28. 475–537.

Bybee, Joan. 1995. Regular morphology and the lexicon. *Language and Cognitive Processes* 10. 425–455.

Jakobson, Roman. 1948. Russian conjugation. *Word* 4. 155–167.

Kholodilova, Maria. 2013. *Pozicionnye svojstva mestoimenij v russkom jazyke* ('Positional characteristics of Russian pronouns). St.Petersburg: St.Petersburg State University, MA thesis.

Knjazev, Jurij. 2007. *Grammatičeskaja semantika: Russkij jazyk v tipologičeskoj perspective* ('Grammatical semantics: Russian language in a typological perspective'). Moscow: Jazyki Slavjanskix Kul'tur.

Lyashevskaja, Olga & Sergej Sharoff. 2009. *Častotnyj slovar' sovremennogo russkogo jazyka* ('Frequency dictionary of Modern Russian'). Moscow: Azbukovnik.

Magomedova, Varvara. 2013. *Lingui-Pingui: a script collection for linguistic search.* St.Petersburg: St.Petersburg State University, ms. Online: https://linguistics.stonybrook.edu/students/varvara.magomedova (4 July, 2016.)

McCarthy, John. 2005. Optimal paradigms. In Laura J. Downing, Tracy Alan Hall and Renate Raffelsiefen (eds.), *Paradigms in Phonological Theory*, 170–210. Oxford: Oxford University Press.

Nesset, Tore & Julia Kuznetsova. 2011. Stability and complexity: Russian suffix shift over time. *Scando-Slavica* 57. 268–289.

Nesset, Tore & Laura A. Janda. 2010. Paradigm structure: Evidence from Russian suffix shift. *Cognitive Linguistics* 21. 699–725.

Shvedova, N. (ed.). 1980. *Russkaja grammatika* ('Russian grammar'). Moscow: Nauka.

Slioussar, Natalia & Maria Kholodilova. 2013. Paradigm leveling in non-standard Russian. In Alexander Podobryaev (ed.), *Formal Approaches to Slavic Linguistics: The Second MIT Meeting 2011*, 243–258. Ann Arbor, MI: Michigan Slavic Publications.

Zaliznyak, Andrej. 1977. *Grammatičeskij slovar' russkogo jazyka. Slovoizmenenie* ('The grammatical dictionary of the Russian language. Inflection'). Moscow: Russkij Jazyk.

PART 2

CHAPTER 6

The Zipfian Paradigm Cell Filling Problem

James P. Blevins, Petar Milin, and Michael Ramscar

1 Introduction

This chapter outlines a general view of form variation as reflecting different states of equilibrium between competing communicative pressures. Two dominant pressures are considered here. The first is a pressure to *discriminate* forms and the messages they express, which has the effect of enhancing differences between expressions. This pressure pushes forms towards the discriminative patterning expressed by the 'one form-one meaning principle'. Unchecked, this pressure can in principle lead to suppletion of the kind reported in languages such as Yélî Dnye (Henderson 1995). However, in most languages, the pressure towards maximal discrimination is countered by a second pressure, which favours regular patterns of form and distribution that facilitate the *prediction* of unencountered forms. The trade-off between these competing pressures helps to account for the coexistence of patterns exhibiting varying degrees of discriminability and regularity in many languages.

These pressures interact with distributional factors in characteristically different ways. Highly irregular formations, such as cases of suppletion, must maintain a sufficiently high *token frequency* in order to remain part of a language. There is likewise a *type frequency* threshold that patterns must maintain in order to qualify as regular. These requirements follow essentially a matter of definition. But regular items (i.e., members of classes with a high type frequency) also raise empirical questions about the *attestation* of variants. It has long been observed that corpora provide only partial coverage of the forms of a language, and that items with large inflectional paradigms tend to be sparsely attested. This chapter presents evidence that the shortfall is greater than previously appreciated, and that the coverage of form variation remains sparse in corpora regardless of size. Corpora obey Zipf's law at all sample sizes, containing a Large Number of Rare Events (LNRE; see Baayen 2000).

The chapter suggests that lexical neighbourhoods play a useful role in defining the classes that support extrapolation from partial samples of inflected variants. The fact that most paradigms are only partially attested inhibits the assignment of items to classes based on patterns of congruent variation across

 | DOI 10.1163/9789004342934_008

cells. Lexical neighbourhoods can help to bootstrap the process of class assignment by defining an initial clustering of items. The deductive value of this clustering derives from a strong correlation, discussed in Section 2.1 below, between similarity at the level of form within neigbourhoods and matching patterns of co-filled cells in paradigms. The co-filled cells in a class provide an analogical base for extending patterns exhibited by one member of the class to other, more sparsely attested, items. The reliability of these deductions in turn provides feedback that can guide subsequent class refinement. The challenge posed by input sparsity is met by extrapolating from classes of items that collectively exhaust the variation exhibited by the class and also contain co-filled cells that provide an analogical base for deducing unencountered forms.

1.1 *The Status of Regularity*

It is often assumed that regularity in a linguistic system is a desirable or normative state and that suppletion and other irregularities represent deviations from the uniform patterns that systems (or their speakers) strive to maintain. This conception derives in part from a pair of more general assumptions that underlie Post-Bloomfieldian approaches to linguistic analysis. The first is that recurrence entails redundancy, and the second is that structure implies decomposed representations. On this view, the goal of morphological analysis involves the distillation of patterns into general symbolic statements, schemas or rules that describe the distribution and interpretation of isolable units of form. Descriptions that exhibit recurrent patterns are regarded as deficient on the grounds that they "fall short of scientific compactness" (Bloomfield 1933: 238) or miss "linguistically significant generalizations" (Chomsky 1965: 42ff.) that should be encapsulated in a symbolic rule system.

Both of these basic assumptions operate with notions of 'identity' and 'redundancy' that are defined primarily in terms of orthographic or phonemic transcriptions. Yet the usefulness of transcriptions for capturing the notion of 'identity' relevant for speakers is challenged by studies that probe sub-phonemic contrasts. Acoustic and psychoacoustic investigation of units ranging in size from words to segments have shown that speakers consistently produce and comprehend durational differences and other types of phonetic variation that do not determine phonemic contrasts. At the word level, Gahl (2008) and Drager (2011) found systematic differences in duration between ostensibly homophonous items in English. At the segment level, Plag et al. (2017) found "significant differences in acoustic duration between some morphemic /s/'s and /z/'s and non-morphemic /s/ and /z/".

Similar contrasts distinguish seemingly recurrent morphological units. Davis et al. (2002) reported differences in duration and fundamental frequency between words and morphologically unrelated onset words. In the domain of inflection, Baayen et al. (2003) found that speakers produced Dutch nouns with a longer mean duration when they occurred as singulars than when they occurred as the stem of the corresponding plural. Kemps et al. (2005) subsequently tested speakers' sensitivity to prosodic differences between singular nouns and 'homophonous' stems, and concluded that "acoustic differences exist between uninflected and inflected forms and that listeners are sensitive to them" (Kemps et al. 2005: 441).

In short, many apparently recurrent units are neither identical nor redundant. Hence regularity cannot be treated as an intrinsic 'design feature' of language or even as a simple reflex of a combinatoric economy metric that rewards analyses in which a recurrent element is "noted only once, with a full statement as to where it is and where it is not used" (Bloomfield 1933: 238). The pervasive regularity observed in languages cannot simply be taken for granted but stands in need of explanation.

A general explanation for regularity should not only identify the factors that are favourable to the creation of regular patterns but also clarify how the function of these patterns contributes to their persistence. At the same time, these considerations should be compatible with the ubiquity of irregularity. Exceptionless regularity is comparatively rare outside artificial languages like Esperanto and even descriptions of standard languages tend to underestimate the amount of variation.

As with regularity, a general account of irregularity should identify the sources of irregular formations and also clarify their functions. Traditions that treat regularity as normative tend to regard irregularity as functionless 'historical residue' or 'noise'. Individual approaches within these traditions develop formal strategies for accommodating deviations from regular patterns, ranging from lexical lookup routines, through 'readjustment rules' to 'default overrides'. However these strategies act essentially as correctives to a conception of language structure as normatively regular and thus offer no insight into the persistence and ubiquity of irregularity.

1.2 *Discriminability and Sparsity*

From a discriminative learning perspective of the kind developed Ramscar & Yarlett (2007); Ramscar et al. (2010) and Ramscar (2013), the situation is reversed. Suppletive forms and other types of irregular formations tend to be maximally discriminated, establishing highly transparent form/meaning contrasts. To the extent that irregular patterns enhance the discriminability of forms, they contribute to the communicative efficiency of a language. In the discriminative model of Ramscar et al. (2013), the main difference between overtly suppletive forms such as *mouse/mice* and more regular forms such as *rat/rats* is that the former serve to accelerate the rate at which a speakers' representation of a specific form/meaning contrast becomes discriminated from the form classes that express similar contrasts. Thus learning serves to increase the general level of suppletion in form-meaning mappings.

It is thus *regularity* that stands in need of explanation in a discriminative approach. Models of grammaticalization suggest a source for regular formations in the morphologization of syntactic configurations. It remains then to account for the persistence and function of regular patterns within a morphological system. One explanation can be found in the structure of the linguistic input. Previous debates regarding the poverty of the stimulus have mainly been concerned with phenomena in the syntactic domain, where there appear to be no cases that withstand serious scrutiny (see, e.g., Pullum & Scholtz 2002; Clark & Lappin 2011). However, the problem of input sparsity arises in an acute form in the morphological domain, given the distributional biases of the forms of a language. These biases reflect Zipf's law (Zipf 1935, 1949), according to which the frequency of a word in a corpus is inversely proportional to its rank in the corpus. As language samples increase in size, they reinforce the rank-size distributions established in smaller samples.[1]

As Kurumada et al. (2013: 440) note, "while Zipfian distributions are ubiquitous across natural language, their consequences for learning are only beginning to be explored". A consequence of immediate morphological relevance is that speakers must learn a language from a partial and biased sample. Although it is generally accepted that exposure to specialized vocabulary and archaic formations may vary across individuals, it is often implicitly assumed that speakers encounter the majority of regular forms in their language. A related assumption underlies the different ways that inflectional and derivational patterns have been investigated in psycholinguistic studies. Since derivational processes are known to exhibit a high degree of item-specific variation, derivational families (de Jong et al. 2000; Mulder et al. 2014) are defined in terms of type counts. In contrast, inflectional processes are assumed to be highly productive, defining uniform paradigms within a given class. Lemma size is not expected to vary, except where forms are unavailable due to paradigm 'gaps' or 'defectiveness'. This allowed studies of inflection to abstract away from variation in type counts and focus on token counts (Baayen et al. 1997; Hay 2001).

1 This chapter is concerned solely with the effects of the patterns described by the laws attributed to Zipf and Herdan (Herdan 1960). For discussion of the source of these patterns, see Ramscar (2017).

Yet studies of corpora, which provide the best available model of language input, indicate that lemma family size also varies considerably. The linguistic significance of this variation is suggested by the fact that lemma family size is a useful predictor of derivational family size, as shown in Milin et al. (2013). But what appears to be clear in any case is that many potentially available inflected forms are unattested in corpora. As corpora increase in size, they do not converge on uniformly populated paradigms. Instead, they reinforce previously attested forms and classes while introducing progressively fewer new items. This distribution suggests that inflected variants of open-class items obey Zipf's law at all observed sample sizes.

The observation that speakers never encounter all of the inflected forms of their language entails that they must be able to solve what Ackerman et al. (2009) term the 'Paradigm Cell Filling Problem' on the basis of the forms that they do encounter. Regularity contributes to this solution by facilitating prediction from a partial and biased sample. From a learning-based perspective, regularity and irregularity are best understood in terms of the complementary functions they serve within a system. Regularity is not normative; it is merely the prerequisite for prediction from Zipf-distributed input. An increase in the discriminability of forms and contrasts aids communicative efficiency. A key point about irregular forms from this perspective is that they are *frequent*—they are in the head of the Zipfian distribution. Accordingly, they can be expected to be acquired and fully discriminated early (cf. *mouses* / *mice*), before the prefrontal cortex develops fully (cf. Ramscar & Gitcho 2007) so that learning is not influenced by the top-down factors that often inhibit the acquisition of irregulars in adult learners. From this perspective, irregulars would exemplify the 'end point' of language learning (as well-discriminated, suppletive forms), but because language is Zipf-discriminated, there is no 'end point' of language learning, and so the distribution needs to be predictable for the tail.

Thus the coexistence of regular patterns and irregular forms is not due solely to the inertia of functionless historical residue but reflects the interaction of competing discriminative and predictive pressures. Once established, irregular formations function as highly-discriminated exponents of properties and as attractors that enhance the salience of regular contrasts. Yet increases in the discriminability of irregulars are offset by a reduction in predictive value. The structure of the input imposes limits on how irregular a language can become without sacrificing learnability.

1.3 *It Takes a Neighbourhood*

In order for a collection of partial samples to guide the prediction of unattested forms, the forms that speakers do know must be organized in such a way that they collectively exhaust the inflectional variation in a language. This gives rise to what, adapting Ackerman et al. (2009), might be called the 'Paradigm Cell Alignment Problem' (PCAP), since speakers must recognize which sets of inflected forms can be 'pooled' to cover the variation that characterizes each class of paradigms. As clearly recognized by Hockett (1967), this problem is far more pressing for speakers than the pedagogical problem of selecting 'principal parts' or 'exemplary paradigms'.

> in his analogizing ... [t]he native user of the language ... operates in terms of all sorts of internally stored paradigms, many of them doubtless only partial; and he may first encounter a new basic verb in any of its inflected forms. (Hockett 1967: 221)

We suggest that the organization required to solve the PCAP is provided by lexical *neigbourhoods* (Baayen et al. 2006; Gahl et al. 2011). Neighbourhoods bootstrap the creative engine of the morphological system by seeding a class structure that permits the analogical deduction of the full system from collections of partial paradigms with systematic patterns of co-filled cells. The role assigned here to neighbourhoods thus builds on the results reported in Milin et al. (2011). In this study, analogical extrapolation from a small set of nearest neighbors allowed a system to model the choice of masculine instrumental singular allomorph by Serbian speakers presented with nonce words. More generally, regular paradigms enable speakers to generate previously unencountered forms, not by appealing to an explicit rule, or to any kind of explicit grammatical knowledge, but because they are implicit in the distribution of forms and meanings in the language as a system.

The following section outlines empirical evidence that Zipf's law determines input sparsity of inflectional variants at all sample sizes and suggests how the structure of lexical neighbourhoods facilitates prediction from sparse input. A central hypothesis explored in this study is that the structure of morphological systems does not reflect the types of formal constraints on representations or rule systems proposed within theoretical models, but derives instead from three primary factors. The first is the Zipfian structure of the input that speakers are exposed to. The second is the discriminative learning strategies that speakers employ when exposed to that input. The third is the structure of lexical neighbourhoods. From this learning-based perspective, the structure of morphological systems is not anchored in any kind of formal architecture or 'innate language faculty' but emerges mainly from the distributional biases of the forms in the system and the general-purpose discriminative learning strategies and principles of analogical deduction employed by speakers.

2 The Zipfian Paradigm Cell Filling Problem

It might seem reasonable to expect that all or most of the inflectional variants of the regular items in a language would eventually show up in the input encountered by speakers, given a large enough sample. From this perspective, one might expect an initial spike of forms with high token frequency that would gradually give way to a more uniform distribution as sample size increases. Yet this is not at all what we find if we examine the distribution of forms in corpora. Figure 6.1 displays rank orders in random samples of the SdeWaC corpus (Faß & Eckart 2013) of German.

The samples in Figure 6.1 start with 1 million distinct word forms and increase stepwise to to 15 million forms, at which point the 850-million word corpus is essentially exhausted. At each size increment, the Zipfian structure becomes more, not less, pronounced, as the head of the distribution grows faster than its tail.

Figure 6.2 exhibits the result of applying a similar sampling methodology to inflected noun variants in the SdeWaC corpus. The growth in the distributional bias of nouns shows same pattern as the randomly-sampled words in Figure 6.1. As sample size increases, the average number of attested inflected noun variants decreases.

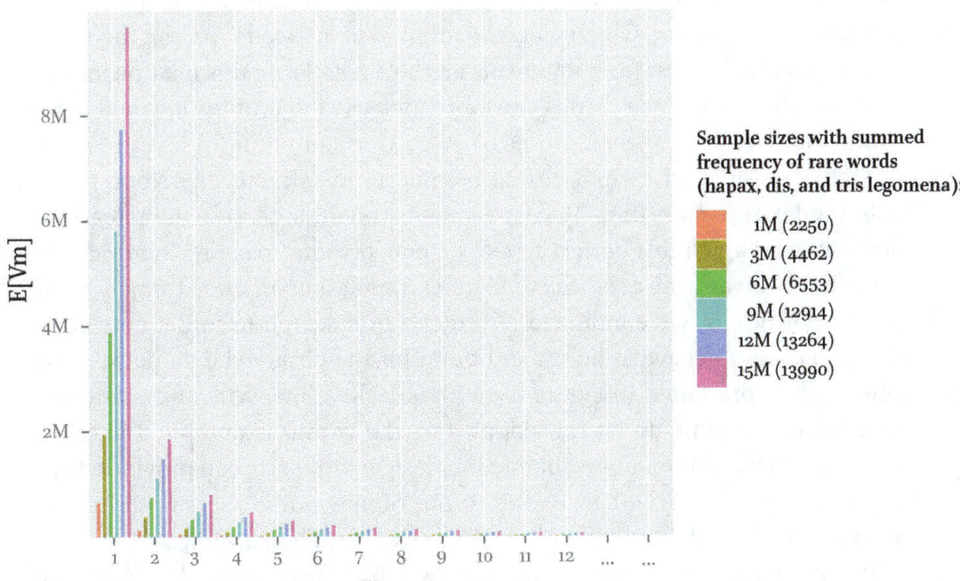

FIGURE 6.1 *Zipf plot for randomly sampled words.*

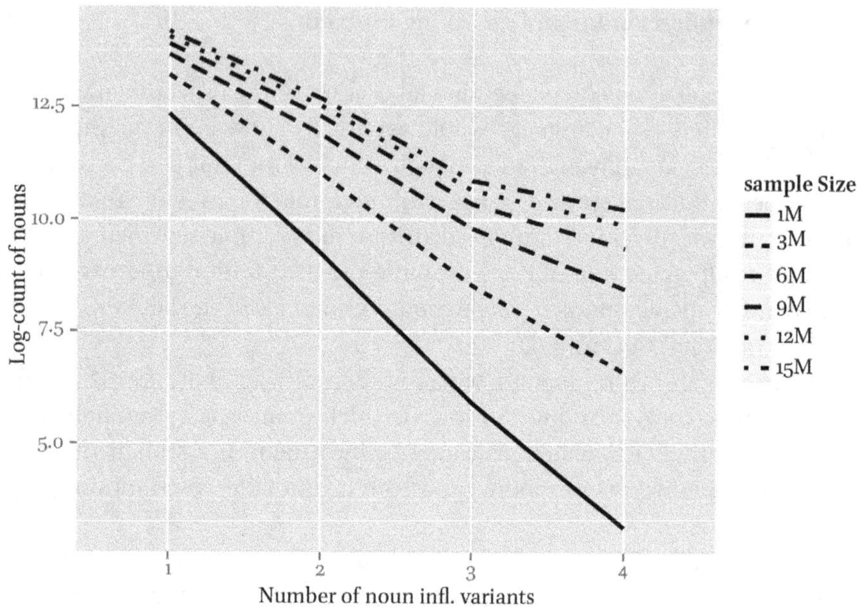

FIGURE 6.2 *The paradigm non-filling pattern.*

As sample size increases, there is also a marked attenuation in the steepness of the slope, though it never becomes completely flat. This trend is extracted and presented in Figure 6.3, which plots the number of attested forms on the X-axis and slopes of six trends from Figure 6.2 on the Y-axis. From this relationship we can infer that even if the corpus size were increased to infinity, it would never contain all possible inflected forms of every German noun. To test this claim explicitly we applied a curve-fitting technique to estimate the slope for an unlimited corpus size (lim(X)). An inverse-exponential three-parameter function obtained an almost perfect fit ($R^2 = 0.996$; p-value <0.0001). To bring even greater conservativeness, since we were extrapolating on an extremely small number of points, we conducted a numeric grid-search to find a new curve that will always behave as an upper bound for the observed data points. This new curve is presented in Figure 6.3 with a dashed line. With this additional restriction, the estimate for the slope when the corpus is unlimited in size is $Y = -1.326$. What this means is that no increase in input size can provide a solution to the PCFP. Even with an endless influx of new words, there would always be more nouns with a smaller number of attested inflected variants.

In summary, not only do the forms drawn from this corpus obey Zipf's law at all sample sizes but the rank-order biases become stronger as sample size increases. To the extent that corpora provide a reasonable approximation of input, these patterns suggest speakers must be able to acquire their language

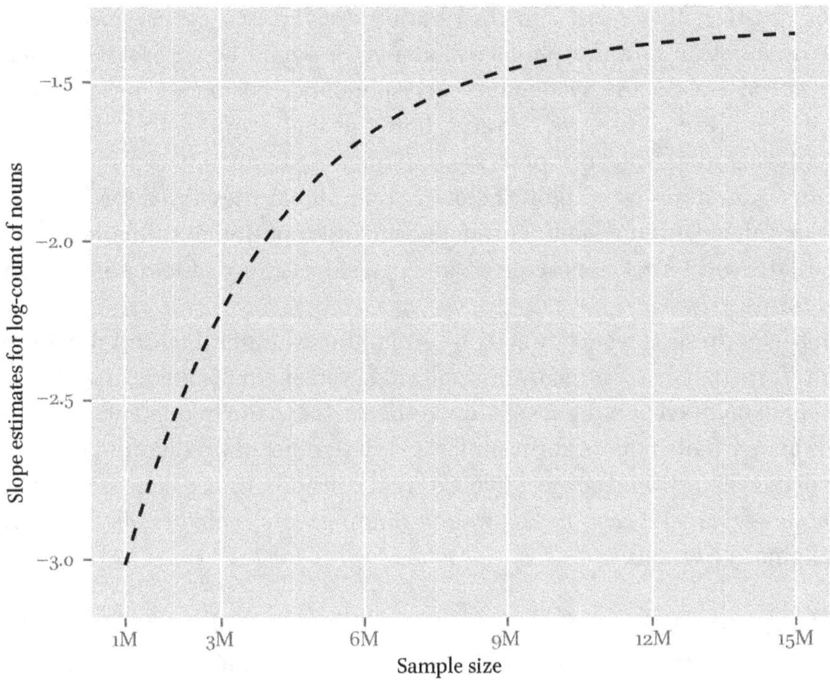

FIGURE 6.3 *You can't get there from here: Asymptoting slopes.*

from exposure that will always be partial, and will in some respects become sparser with increased exposure.

2.1 *Formal Similarity and Paradigmatic Analogy*

The Zipfian distribution exhibited in Figures 6.1–6.3 is also characteristic of the forms that fill cells of inflectional paradigms (see, e.g., Kostić et al. 2008). The exact distribution may of course be to some degree item-specific: there is no reason to assume that the probability-based rank-order must be the same for the paradigm cells of all words. A number of different factors may influence which cells of a given paradigm are filled. Semantic considerations will play a role in some cases. For example, the physical properties of the objects associated with nouns are known to influence the probability or even possibility of particular locative or instrumental case forms.

On a Post-Bloomfieldian conception, a language is viewed as an inventory of forms composed of recurrent parts and paradigms are treated as 'epiphenomena'. Hence there are no obvious expectations about patterns of 'paradigm cell filling' apart from any effects attributable to frequency or other independent factors. In contrast, a conception of language as a complex adaptive system leads one naturally to search for functions or other correlates of distributional patterns. Given that form and distribution are the two observable dimensions of variation in a morphological system, it would be surprising if there were not at least some systematic correspondences between these properties. In the present case, we consider how the shape and distribution of attested forms might aid the process of deducing unencountered forms. The specific hypothesis we explore below is that similarities in shape that define lexical neighbourhoods correlate with distributional similarities that permit attested forms to provide an analogical base for deducing unencountered forms.

This hypothesis is tested directly by evaluating whether the form proximity of wordform pairs, measured by normalized Levenshtein distance, predicts the number of co-filled paradigm cells, once effects due to the frequencies of those words are partialled out. If form similarity is a significant predictor of co-filled paradigm cells, it would suggest that external considerations (such as meaning or frequency) are not the sole factors determining which cells are filled in noun paradigms in German.

2.1.1 Methodology

To test this correlation, we turned again to the SdeWaC corpus to sample German noun pairs across frequency bands. Our four-stage method represents a variant of a stratified sampling procedure in which we secured a random fraction of nouns from all frequency strata (the exact procedure was adopted from Ellis & Hooper (2001)). We first selected all nouns with a frequency of 5 or higher. We then calculated the total number of tokens (or the total summed frequency), and arbitrarily set the sample size to 50,000 nouns. The total number of tokens divided by the sample size determined the step size (i.e., the summed frequency per band). In the next stage, nouns were ordered by frequency, and ties were randomized (shuffled). In the final stage we started sampling, from low to high frequency nouns: once the cumulative sum of frequencies was equal or higher than the step size we terminated sampling for a given band and moved to the next, now using the difference between the current cumulative sum and the step size. This process was repeated until we exhausted the full list of nouns, that is, when our sample size reached 50,000 nouns.

Once the nouns were sampled, we formed all possible pairs and calculated their normalized Levenshtein distance (Levenshtein 1966) and determined the number of co-filled paradigm cells (ranging from 0 to 8, reflecting the space defined by the 4 cases and 2 numbers in the German declension system). Since we were interested in the predictability of the closest form neighbours, we retained only those pairs whose normalized Levenshtein distance was less than or equal to 0.5 (i.e., values ranging from 0, indicating no difference, to 0.5, indicating a difference in roughly half of the letters). Our final sample consisted of approximately 27.2 million noun pairs.

To test our prediction, we analyzed our dataset using the generalized additive mixed model (GAMM), in the R statistical environment (R Core Team 2014), with the MGCV package (Wood 2006). A GAMM was fitted to the number of co-filled cells, testing the nonlinear numeric interaction of two word frequencies and, additionally, the effect of the normalized Levenshtein distance, allowing only for mild (minimal) non-linearity. The tensor product of two frequencies appeared highly predictive (edf = 23.834; F = 94468; p <0.0001). Crucially, however, over and above this effect we observed a strong effect of Levenshtein distance (edf = 1.986; F = 3297; p <0.0001). This supports the initial hypothesis that form neighborhoods can serve as the basis for the analogical prediction of inflectional variants.

A GAMM was then fitted to the number of co-filled cells, testing the nonlinear numeric interaction of two word frequencies, cosine similarity between contextual vectors of the same two words,[2] and, finally, the effect of the normalized Levenshtein distance. The tensor product of two frequencies appeared highly predictive (edf = 23.749; F = 79595; p <0.0001), and so did cosine similarity (edf = 4.000; F = 138683; p <0.0001). Crucially, however, over and above these two effects we observed a strong main effect of Levenshtein distance (edf = 1.029; F = 5805; p <0.0001). For this effect we where we allowed only for minimal nonlinearity. The partial interaction of cosine similarity with Levenshtein distance also appeared as statistically significant, but to a lesser degree (edf = 15.039; F = 1458; p <0.0001). These results further support the hypothesis that, in German nominal paradigms, form neighborhoods provide a basis for the analogical deduction of variants.

2.1.2 Testing the Predictions

Figures 6.4–6.7 now exhibit the effects described above. Figure 6.4 shows that frequency effects display the expected increase in the number of co-filled cells as a product of the increase of the frequencies of the respective nouns.

Figure 6.5 then shows the nonlinear effect of cosine similarity, revealing a general trend in which semantic similarity lowers the number of co-filled cells. This partial effect takes only negative values, becoming most negative for words that are most similar. The same general pattern applies to form similarity: Figure 6.6 shows that the number of co-filled cells steadily decreases as form similarity decreases.

2 For detailed discussion of the theoretical and technical aspects of vector-based semantic similarity, see Lund & Burgess (1996) and Shaoul & Westbury (2010).

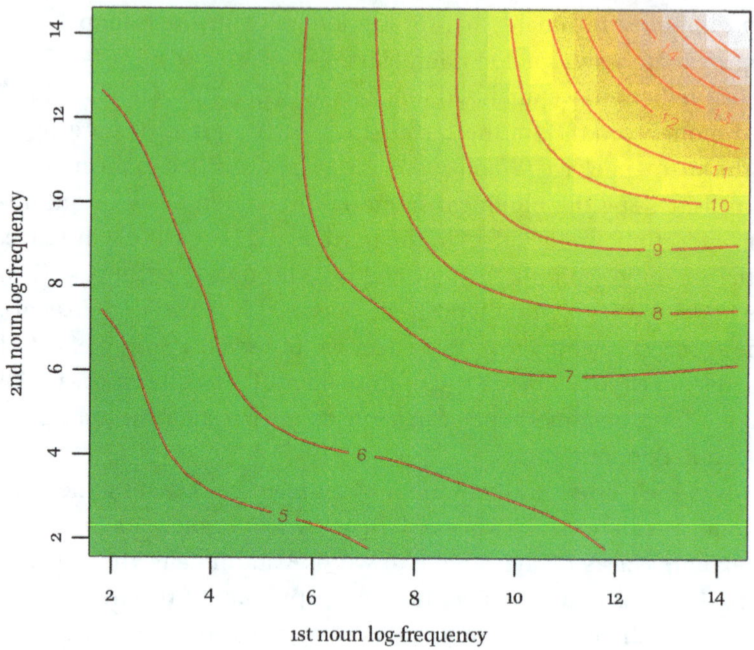

FIGURE 6.4 *Nonlinear interaction (tensor product) effect of frequencies of paired nouns for the number of co-filled paradigm cells.*

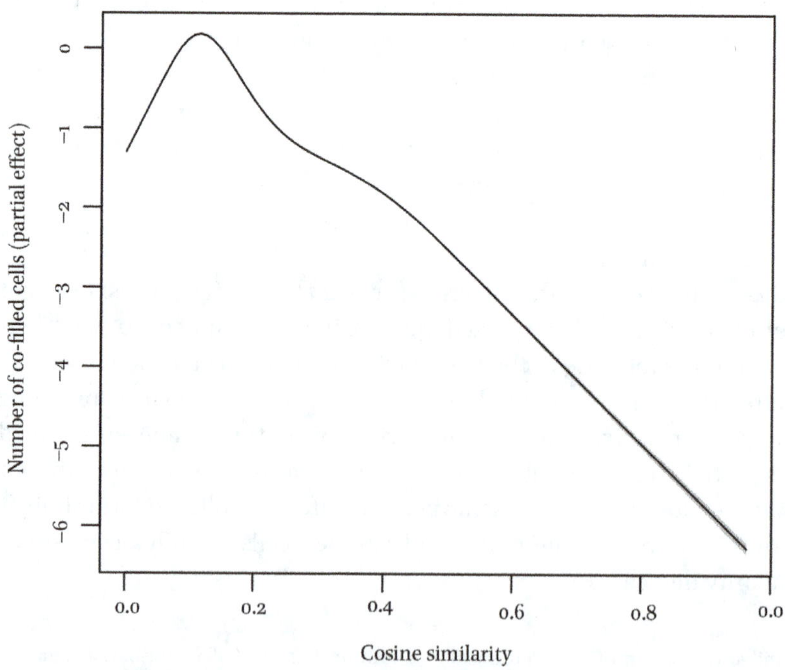

FIGURE 6.5 *Partial effect of the cosine similarity between contextual vectors of two nouns for the number of co-filled paradigm cells.*

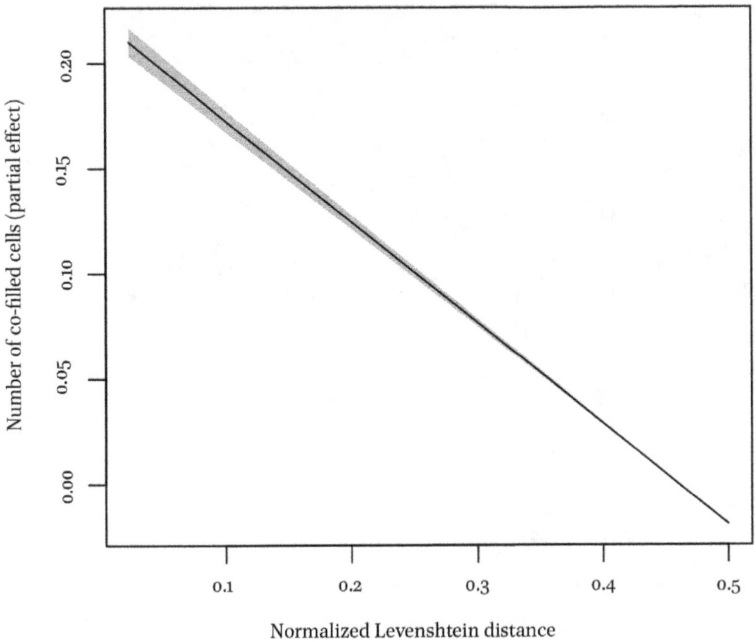

FIGURE 6.6 *Partial effect of the normalized Levenshtein distance between two nouns for the number of co-filled paradigm cells.*

In this case, we see that the partial effect takes only positive values, but is a degree weaker than the effect of cosine similarity displayed in Figure 6.5.

Finally, although the partial interaction between cosine similarity and Levenshtein distance may have appeared to be statistically significant, Figure 6.7 reveals that it is, in fact, not substantial. Figure 6.7 exhibits the same cosine similarity effect, where only for the most similar nouns Levenshtein distance forms a U-shaped plane. However, we can also see that there are no nouns which are very similar in contextual appearance and closest form neighbors at the same time (that is, the upper left quadrant does not contain any values).

2.1.3 Summary

Figures 6.4–6.7 show that the number of co-filled cells steadily decreases as the form similarity decreases. A learning framework provides a simple and elegant explanation for the observed form similarity effect (operationalized by normalized Levenshtein distance). Words attested in larger corpus should manifest analogical predictions when they are similar in form. And exactly that is confirmed by the number of co-filled paradigm cells as co-determined with the form overlap of pairs of words.

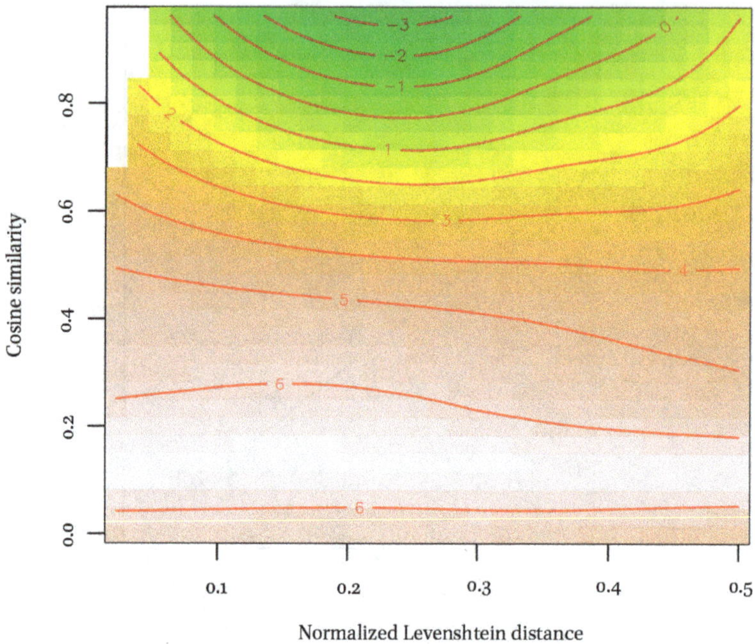

FIGURE 6.7 *Nonlinear interaction (tensor product) effect of cosine similarity and Levenshtein distance between two nouns for the number of co-filled paradigm cells.*

These effects reinforce the previous results of Milin et al. (2011), which showed that the closest form neighbors determine the production of unseen inflected variant of nonce words (in this case an allomorph of masculine instrumental singular). Together, these studies suggest that items similar in form indeed help paradigm cell filling, a conclusion that is also consistent with studies such as Pertsova (2004), which report neighbourhood effects on the inflection of nonce words. The present study demonstrates that the highest cell overlap characterizes the nearest neighbors.

3 A Learning-based Perspective on the Zipfian PCFP

We conclude with a concise summary and a brief discussion of some implications of the results reported above. The point of departure for this study was the proposal that the coexistence of regular and irregular patterns could be understood in terms of a trade-off between the opposing communicative pressures imposed by predictability and discriminability. Somewhat counterintuitively, the pressures associated with irregulars are reasonably transparent, once irregulars are recognized as functional rather than 'defective'. The enhanced discriminability of irregular formations brings them into maximal conformance with precepts like the 'one form-one meaning principle', while their effects as attractors within a larger system are established in studies such as Ramscar et al. (2013). In contrast, recognizing the role that regulars play in enhancing predictability creates a conundrum. Although, by definition, regular items must have high type frequency, the Zipfian distribution of regularly inflected variants entails that many variants have low token frequency or are unattested altogether. In effect, the forms that establish the regularity of regulars must be deduced from partial samples. Section 2.1 suggests how the relation between neighbourhoods and patterns of co-filled cells facilitate this deduction.

The elements of the resulting communicative dynamic are summarized in (1).

(1) THE PREDICTION-DISCRIMINATION DYNAMIC

a. Morphological systems exhibit regularities because, given the Zipfian structure of the input, speakers never encounter all the forms of a language and must be able to predict new forms from partial samples.

b. Irregular formations are persistent because they serve two communicative functions. As individual expressions, they are well discriminated. As exceptional members of larger sets of alternating elements, they emphasize contrasts that are less saliently marked in regular patterns.

c. Lexical neighbourhoods compensate for input sparsity. Although the forms of individual items are partially attested, the inflectional patterns that they follow are robustly attested within their form neighbourhoods.

This initial study also suggests a number of directions for future research. The relation between neighbourhoods and co-filled cells identifies a viable base for analogical extrapolation. Studies such as Ackerman & Malouf (2013) likewise show that paradigms exhibit a degree of mutual informativeness that supports the deduction of unencountered forms. A discriminative learning model is well adapted to exploiting these sources of information in language learning and processing. However, questions remain concerning the cognitive implementation of analogical deduction and the ways that this process interacts with other factors that enter into learning and processing. The amount of cross-linguistic variability in the relation between neighourhood similarity and patterns of co-filled cells also remains an open question. The initial choice of German helps to establish that input sparsity arises in languages with comparatively few inflectional variants. In languages with larger and more variable paradigms, Zipfian biases are expected to create an even greater form shortfall and thereby enhance the deductive value of an analogical base.

A clearer understanding of the way that Zipfian biases constrain the space of solutions for the PCFP is also of value to Post-Bloomfieldian models. To solve the PCFP, combinatoric approaches require a predictive model that overcomes the 'Segmentation Problem' (Spencer 2012) and other challenges to the assumption of lossless decomposition (see, e.g., Blevins 2016). Approaches that aim for psychological relevance also require a notion of 'identity' that takes account of the sub-phonemic variation described in Section 1.1, along with some evidence that the model is learnable from the Zipf-distributed input that speakers can be assumed to encounter.

References

Ackerman, F., Blevins, J.P. & Malouf, R. (2009). Parts and wholes: Implicative patterns in inflectional paradigms. In Blevins, J.P. & Blevins, J. (eds.), *Analogy in Grammar: Form and Acquisition*, Oxford University Press, chapter 3, 54–81.

Ackerman, F. & Malouf, R. (2013). Morphological organization: The Low Conditional Entropy Conjecture. *Language* 89, 429–464.

Baayen, R.H. (2000). *Word Frequency Distributions*. Dordrecht: Kluwer Academic Publishers.

Baayen, R.H., Feldman, L.B. & Schreuder, R. (2006). Morphological influences on the recognition of monosyllabic monomorphemic words. *Journal of Memory and Language* 53, 496–512.

Baayen, R.H., Lieber, R. & Schreuder, R. (1997). The morphological complexity of simple nouns. *Linguistics* 35, 861–877.

Baayen, R.H., McQueen, J.M., Dijkstra, T. & Schreuder, R. (2003). Frequency effects in regular inflectional morphology: Revisiting Dutch plurals. In Baayen, R.H. & Schreuder, R. (eds.), *Morphological Structure in Language Processing*, Berlin: Mouton de Gruyter, 355–370.

Blevins, J.P. (2016). *Word and Paradigm Morphology*. Oxford: Oxford University Press.

Bloomfield, L. (1933). *Language*. Chicago: University of Chicago Press.

Chomsky, N. (1965). *Aspects of the Theory of Syntax*. Cambridge, MA: MIT Press.

Clark, A. & Lappin, S. (2011). *Linguistic Nativism and the Poverty of the Stimulus*. Wiley-Blackwell.

Davis, M., Marslen-Wilson, W.D. & Gaskell, M. (2002). Leading up the lexical garden-path: Segmentation and ambiguity in spoken word recognition. *Journal of Experimental Psychology: Human Perception & Performance* 28, 218–244.

de Jong, N.H., Schreuder, R. & Baayen, R.H. (2000). The morphological family size effect and morphology. *Language and Cognitive Processes* 15, 329–365.

Drager, K.K. (2011). Sociophonetic variation and the lemma. *Journal of Phonetics* **39**, 694–707.

Ellis, N.C. & Hooper, A.M. (2001). Why learning to read is easier in Welsh than in English: Orthographic transparency effects evinced with frequency-matched tests. *Applied Psycholinguistics* **22(4)**, 571–599.

Faß, G. & Eckart, K. (2013). SdeWaC—a corpus of parsable sentences from the web. In Gurevych, I., Biemann, C. & Zesch, T. (eds.), *Language Processing and Knowledge in the Web*, Heidelberg: Springer, 61–68.

Gahl, S. (2008). "Thyme" and "Time" are not homophones. The effect of lemma frequency on word durations in spontaneous speech. *Language* **84(3)**, 474–496.

Gahl, S., Yao, Y. & Johnson, K. (2011). Why reduce? Phonological neighborhood density and phonetic reduction in spontaneous speech. *Journal of Memory and Language* **66(4)**, 789–806.

Hay, J. (2001). Lexical frequency in morphology: Is everything relative? *Linguistics* **39**, 1041–1070.

Henderson, J.E. (1995). *Phonology and Grammar of Yele, Papua New Guinea*. Pacific Linguistics B-112, Camberra: Pacific Linguistics.

Herdan, G. (1960). *Type-Token Mathematics*. The Hague: Mouton.

Hockett, C.F. (1967). The Yawelmani basic verb. *Language* **43**, 208–222.

Kemps, Rachèl J.J.K., Ernestus, M., Schreuder, R. & Baayen, R.H. (2005). Prosodic cues for morphological complexity: The case of Dutch plural nouns. *Memory & Cognition* **33(3)**, 430–446.

Kostić, A., Ilić, S. & Milin, P. (2008). Probability estimate and the optimal text size. *Psihologija* **41(1)**, 35–51.

Kurumada, C., Meylan, S.C. & Frank, M.C. (2013). Zipfian frequency distributions facilitate word segmentation in context. *Cognition* **127**, 439–453.

Levenshtein, V.I. (1966). Binary codes capable of correcting deletions, insertions, and reversals. *Soviet Physics Doklady* **10(8)**, 707–710.

Lund, K. & Burgess, C. (1996). Producing high-dimensional semantic spaces from lexical co-occurrence. *Behavior Research Methods, Instruments, & Computers* **28(2)**, 203–208.

Milin, P., Keuleers, E. & Blevins, J.P. (2013). Lemma size effects on morphological processing. 9th Mediterranean Morphology Meeting, Dubrovnik.

Milin, P., Keuleers, E. & Filipović Đurđević, D. (2011). Allomorphic responses in Serbian pseudo-nouns as a result of analogical learning. *Acta Linguistica Hungarica* **58**, 65–84.

Mulder, K., Dijkstra, T., Schreuder, R. & Baayen, R.H. (2014). Effects of primary and secondary morphological family size in monolingual and bilingual word processing. *Journal of Memory and Language* **72**, 59–84.

Pertsova, K. (2004). *Distribution of Genitive Plural Allomorphs in the Russian Lexicon and in the Internal Grammar of Native Speakers*. Master's thesis, UCLA.

Plag, I., Homann, J. & Kunter, G. (2017). Homophony and morphology: The acoustics of word-final S in English. *Journal of Linguistics* 53(1), 181–216.

Pullum, G.K. & Scholtz, B. (2002). Empirical assessment of stimulus poverty arguments. *The Linguistic Review* 19, 8–50.

R Core Team (2014). *R: A Language and Environment for Statistical Computing.* R Foundation for Statistical Computing, Vienna, Austria.

Ramscar, M. (2013). Suffixing, prefixing, and the functional order of regularities in meaningful strings. *Psihologija* 46(4), 377–396.

Ramscar, M. (2017). The discriminative nature of human communication. ms. Eberhard-Karls-Universität, Tübingen.

Ramscar, M., Dye, M. & McCauley, S.M. (2013). Error and expectation in language learning: The curious absence of *mouses* in adult speech. *Language* 89(4), 760–793.

Ramscar, M. & Gitcho, N. (2007). Developmental change and the nature of learning in childhood. *Trends in Cognitive Sciences* 11(7), 274–279.

Ramscar, M. & Yarlett, D. (2007). Linguistic self-correction in the absence of feedback: A new approach to the logical problem of language acquisition. *Cognitive Science* 31, 927–960.

Ramscar, M., Yarlett, D., Dye, M., Denny, K. & Thorpe, K. (2010). The effects of feature-label-order and their implications for symbolic learning. *Cognitive Science* 34, 909–957.

Shaoul, C. & Westbury, C. (2010). Exploring lexical co-occurrence space using HiDEx. *Behavior Research Methods* 42(2), 393–413.

Spencer, A.J. (2012). Identifying stems. *Word Structure* 5, 88–108.

Wood, S.N. (2006). *Generalized Additive Models: An Introduction with R.* CRC Press.

Zipf, G.K. (1935). *The Psychobiology of Language: An Introduction to Dynamic Philology.* Cambridge, MA: MIT Press.

Zipf, G.K. (1949). *Human Behavior and the Principle of Least Effort.* Cambridge, MA: Addison-Wesley.

CHAPTER 7

Paradigm Realisation and the Lexicon*

Géraldine Walther

Introduction

Morphology is defined as the study of word forms. In a Word and Paradigm (WP) approach (Hockett 1954; Robins 1959; Matthews 1972, 1974; Ackerman et al. 2011; Blevins 2016, among others), this includes the study of individual word forms as well as the study of the paradigmatic properties which shape the overall morphological system. Aronoff (1994) illustrates how some of those properties define a specific domain of organization, which he calls the MORPHOMIC LEVEL: morphomic level properties are shared by multiple lexical items within a language and appear to express contrasts that are neither fully reducible to phonological word properties nor to syntactic relationships between individual words. Within the domain of inflectional morphology, those morphomic properties correspond to a lexeme's paradigm structure and constitute the core domain of investigation of WP approaches.

Despite the fundamentally lexical dimension of morphology defined in these terms, WP approaches falling into the realizational family (Matthews 1965; Anderson 1982; Stump 2001; Spencer 2005, 2013; Stump 2006; Brown and Hippisley 2012; Crysmann and Bonami 2015, among others) concentrate on the structure of morphological forms and the information needed to account for all word forms and complete paradigm realization. Yet they seldom offer a fully explicit definition of the lexical entries whose paradigmatic properties are being investigated. The same is in most cases true for the great variety of subtle differences between individual paradigms, which can only be accounted for through large scale data analysis, if at all. But, with possibly the exception of

* **Acknowledgements:** For comments on previous versions of this paper and/or fruitful discussions leading up to the development of PARSLI, the author would like to thank Olivier Bonami, Greg Stump, Dunstan Brown, Jim Blevins, and especially Benoît Sagot. The major part of this paper was written while the author was affiliated with the CNRS research unit Dynamique du Langage. The author is grateful to the ASLAN project (ANR-10-LABX-0081) of the Université de Lyon, for its financial support within the program "Investissements d'Avenir" (ANR-11-IDEX-0007) of the French government operated by the National Research Agency (ANR).

 | DOI 10.1163/9789004342934_009

Network Morphology (Brown and Hippisley, 2012), most approaches are only peripherally associated with implementation tools that would permit investigations of the considerable variation across items in an inflectional system. Even implemented analyses mostly cover small or, at best, medium scale subsets of the lexicon in any of these approaches. It is arguably only through large scale analyses that the broad range and distribution of variation between individual lexical items can be revealed, as the overall morphological system is shaped by the differences between its individual components and the proportion of individual entries conforming to specific patterns. In order to access this broad range of variation, it seems crucial to shift part of our focus towards the individual lexical properties of morphological systems.

The aim of the present paper is to concentrate more narrowly upon the definition of lexical information as relevant to defining the paradigmatic properties of individual lexemes on the one hand and the properties of the lexicon of a language on the other. As a way of representing the lexical properties of morphological systems, this paper presents PARSLI, a model of the relationship between lexical properties and inflectional paradigmatic properties. This model in particular focuses on representing the locus of variation between paradigms from a realizational perspective.

All versions of the PARSLI model have been equipped with a dedicated implementation, Alexina PARSLI (Sagot and Walther, 2013) based on the Alexina lexical framework (Sagot, 2010) and allowing for large scale implementations of grammatical descriptions. Alexina_PARSLI has been described in (Sagot and Walther, 2013; Walther, 2013). Formal details thereof can be found in these works.

Lexical Entries in WP Approaches

The Lexicalist Heritage

As illustrated by Ackerman, Stump, and Webelhuth (2011), realizational approaches to inflectional morphology are firmly embedded in a lexicalist approach to grammar, most prominently represented by Lexical Functional Grammar (LFG: Bresnan 1982; Dalrymple 2001) or Head-Driven Phrase Structure Grammar (HPSG: Pollard and Sag 1994). In particular, as characterized by Ackerman et al. (2011), approaches such as Network Morphology (NM: Corbett and Fraser 1993; Brown and Hippisley 2012), Paradigm Function Morphology (PFM: Stump 2001), Generalised Paradigm Function Morphology (GPFM: Spencer 2005, 2013), Realization-based Lexicalism (RbM: Ackerman et al. 2011), or classical to contemporary HPSG-based approaches to morphology (from Koenig 1999 to Information Based Morphology (IBM) by Crysmann and Bonami 2015) comply with at least three if not all four of the following principles:

1. Principle of morphological integrity
Syntactic mechanisms neither make reference to a word form's proper subparts nor can they create new word forms in a constituent structure.
2. Principle of lexical modification
The lexical properties (meaning, argument structure, grammatical function inventories, and case government patterns) associated with a lexeme are fully determined by lexical stipulation together with rules of lexeme derivation and cannot be altered by items of the syntactic context in which a realization of that lexeme appears.
3. Principle of morpholexical inflection
The morphosyntactic content associated with a lexeme's realizations is fully determined by lexical stipulation together with rules of inflectional morphology and cannot be altered by items of the syntactic context in which a realization appears.
4. Principle of unary expression[1]
In syntax, a lexeme is uniformly expressed as a single morphophonologically integrated and syntactically atomic word form.

ACKERMAN ET AL., 2011

Interestingly, most existing realizational approaches focus almost exclusively on the elaboration of a particular rule architecture and the mapping of morphosyntactic property sets to sets of word forms. These elaborations specify the definition of specific rules, alternative strategies for determining the internal or external ordering of those rules, the types and nature of feature/form correspondences. There is a particular focus on form/feature mismatches, syncretisms, and other irregularities, as well as the status of rule-based feature relations with respect to these irregularities, and the definition of principle parts as minimum sets of forms whose stems are sufficient to realize all paradigm forms.

These factors tend to leave the definition of the interface between inflectional paradigms and the nature of the lexicon largely, if not completely, implicit. There is a striking contrast with syntax, where the the explicit definition

1 This last principle is mentioned with respect to some realizational approaches by Ackerman et al. (2011) but it is not considered a necessary one by the authors. In particular, it does not apply to their own RbM.

of lexical entries constitutes a major focus in lexicalist approaches. Developing an equally explicit definition of lexical entries within realizational models of morphology would thus constitute a promising strategy for bridging work on morphology and syntax, while allowing for a full integration of existing syntactic and morphological frameworks.

Although inflectional, and more specifically paradigmatic, irregularity has been an important focus of recent realizational approaches,[2] little work so far has specifically addressed the nature of a realizational lexicon and the representation of those irregularities within inflectional lexical entries. Thus the defined architecture is usually not built on an explicitly specified theory of the lexicon that would ground a given realizational approach in a lexicalist approach to grammar.[3]

The model presented in this paper aims to address this deficiency by providing an explicit definition of realizational lexical entries that could be integrated in various lexicalist models of grammar. The present model focuses on the relationship between paradigmatic structures in a language and the properties of a language's individual lexical entries. In particular, it highlights the origin of lexeme-specific inflectional irregularities, and offers a precise, lexically integrated, formalized, and quantifiable definition of fundamental inflectional notions such as INFLECTIONAL CLASSES, INFLECTIONAL CATEGORIES, MORPHOSYNTACTIC FEATURES, and INFLECTIONAL REGULARITY.[4]

---

2 See for example the extensive work on irregularity carried out within the paradigm of Canonical Typology (CT: Corbett 2007; Brown and Chumakina 2012, among others).

3 IBM (Crysmann and Bonami, 2015), being fully integrated into and built out of HPSG, does however present a natural exception to this tendency: morphology as part of lexical entries is integrated into the general HPSG sign hierarchy, and thus specified as such. The main focus within current work in this framework however is still on the definition of rules, and, more specifically, on the definition of separate exponent realization and morphotactics—as also proposed in GPFM by Spencer (2013) and more recently in an extension of PFM (Stump, 2014).

4 PARSLI has originally been developed for precise formalization of regularity notions developed within the typological framework of CT. One aim was to explicitly model the regularities and irregularities within a paradigm and/or an inflectional system. The irregularities that are represented within the non-standard lexical entries described hereafter formally represent the so called NON-CANONICAL PHENOMENA within canonical typology. Among those are suppletion (Boyé, 2006) and stem alternations, heteroclisis (Stump, 2006), deponency (or morphosyntactic mismatches) Baerman (2007a), defectiveness (Baerman et al., 2010), overabundance (Thornton, 2011), deficiency and overdifferentiation (Walther, 2013), and syncretisms. Each non-canonical phenomenon is represented as pieces of information directly encoded within the structure of a specific lexical entry. Moreover, the extent of the non-canonical phenomenon can be quantified with specific non-canonicity measures developed

Representations of Lexical Entries: A Quick Overview

Even though the lexicon has not been their primary focus, existing realizational frameworks, from Matthews (1965) to PFM (Stump, 2001), GPFM (Spencer, 2013), or NM (Corbett and Fraser, 1993; Brown and Hippisley, 2012), have all discussed, and more or less specifically detailed, the dimensions defining lexical entries, or lexemes.

Lexical Entries in (Matthews, 1965)

Matthews' initial representation, inherited from the definition of a LEXEME proposed by Lyons (1963, p.12), characterizes a lexeme along two primary dimensions:

> These [lexemes—GW] will contain two relevant sorts of information. In the first place, they will specify that the root of LOVE is *love*, that of SING *sing*, and so forth. Secondly, they will list the MORPHOLOGICAL CLASSES to which each lexeme is assigned: thus the entry for SING will indicate that SING is a member of A, the entry for Latin SAPIENS that SAPIENS is a member of B, and so forth.
>
> MATTHEWS, 1965, P. 145

This definition already comprises the core dimensions that have been developed in later approaches: stem inventories and inflectional classes.

Lexical Entries in PFM (Stump, 2001, 2006)

In PFM (Stump, 2001, 2006) lexical entries are defined along three dimensions: a set of stems, inflectional class membership, and a set of morphosyntactic properties that can be realized through the application of rules.

The set of morphosyntactic properties forms a lexeme's CONTENT PARADIGM, that is a set of opaque ordered pairs $<L,\sigma>$, where L is the lexeme identifier and σ one of the lexeme's expressed morphosyntactic property sets. Through what is called a rule of PARADIGM LINKAGE, the lexeme's content paradigm is linked to the the lexeme's FORM PARADIGM comprising couples $<s,\tau>$, where s is one of L's (lexically listed) stems and τ is the realized morphosyntactic property set. The default link between the two paradigms is expressed through what Stump (2006) calls the UNIVERSAL DEFAULT RULE OF PARADIGM LINKAGE (Stump, 2006, p. 286). In the default case τ in $<s,\tau>$ is identical to σ in $<L,\sigma>$.

---

within the framework (Walther, 2013), which, for lack of space, won't be further discussed in this paper.

Figure 7.1 shows an example of the content and form paradigms for a Latin noun like VERBUM word whose full paradigm is given in Table 7.1. They conform to the default rule of paradigm linkage.

The word forms filling the cells of a lexeme's content paradigm are realized using the appropriate realization rule. Realization rules are of the form given in (1), where *f* is a realization rule applied to a morphological stem X of class C realizing the morphosyntactic properties σ.

(1) $X_C\sigma \rightarrow f(X)$

Content paradigm of VERBUM:	Form paradigm of VERBUM:
<VERBUM, {NOM, SG}>	<verb, {NOM, SG}>
<VERBUM, {VOC, SG}>	<verb, {VOC, SG}>
<VERBUM, {ACC, SG}>	<verb, {ACC, SG}>
<VERBUM, {GEN, SG}>	<verb, {GEN, SG}>
<VERBUM, {DAT, SG}>	<verb, {DAT, SG}>
<VERBUM, {ABL, SG}>	<verb, {ABL, SG}>
<VERBUM, {NOM, PL}>	<verb, {NOM, PL}>
<VERBUM, {VOC, PL}>	<verb, {VOC, PL}>
<VERBUM, {ACC, PL}>	<verb, {ACC, PL}>
<VERBUM, {GEN, PL}>	<verb, {GEN, PL}>
<VERBUM, {DAT, PL}>	<verb, {DAT, PL}>
<VERBUM, {ABL, PL}>	<verb, {ABL, PL}>

FIGURE 7.1 *PFM content and form paradigms for the Latin noun VERBUM 'word'.*

TABLE 7.1 *Latin paradigms for AMĪCUS 'friend' and VERBUM 'word'*

	SG	PL	SG	PL
NOM	amīcus	amīcī	verbum	verba
VOC	amīce	amīcī	verbum	verba
ACC	amīcum	amīcōs	verbum	verba
GEN	amīcī	amīcōrum	verbī	verbōrum
DAT	amīcō	amīcīs	verbō	verbīs
ABL	amīcō	amīcīs	verbō	verbīs

Realizing the <verb, {GEN, PL}> cell in VERBUM's form paradigm thus corresponds to applying the realization rule in (2) which is valid for o-stem nouns (Cl-o).[5]

(2) $X_{\text{Cl-o}}$, {gen, pl} → X ōrum

Inflectional class membership is built into a more general hierarchy. For example, the Latin nouns AMĪCUS 'friend' and VERBUM 'word', whose paradigms are exemplified in Table 7.1, both belong to the class Cl-o. They do however present a few differences in their nominative, accusative plural, and vocative singular forms. A way to represent those differences is to stipulate that they belong to two subclasses of a common noun inflection class. Every member of a subclass inherits all the properties of its superclass. Thus applicable realization rules for a given lexeme can make reference to either the more general superclass or one of its subclasses. Thus the realization rule in (2) applies to both AMĪCUS and VERBUM, but the rule in (3) only applies to AMICŪS while (4) only applies to VERBUM.

(3) $X_{\text{Cl-o1}}$, {acc, pl} → X ōs

(4) $X_{\text{Cl-o2}}$, {acc, pl} → X a

It is however telling that the three dimensions that would identify lexical entries in PFM (stems, inflection class, and sets of morphosyntactic properties) are reflected most directly in the form of PFM realization rules and do not directly define any such entry: realization rules are associated with lexemes, according to their class membership and applied to (one of) their stem(s), as illustrated in (1). As such, PFM provides a perfect illustration of how the definition of lexical entries has been a secondary concern in contemporary realizational WP morphology.

Lexical Entries in GPFM (Spencer, 2005, 2013)

In GPFM (Spencer, 2013), form realization is handled in a similar fashion. Yet, GPFM, which is distinguished in part by the extension of PFM to derivational morphology, offers extended lexical entries. Their definition comprises four dimensions: a morphological representation (MOR), a syntactic representation

5 In the case of form realization through multiple, stacked, realization rules, the result of the application of one rule becomes the morphological stem, that is the input, of the following rule.

	Stem0	*draw*
MOR(87,98)	Stem1	*drew* 'Past'
	Stem2	*drawn* 'Past Participle'
	MorCat	V
SYN(87,98)	V(SUBJ; (OBJ))	
SEM(87)	'MAKE_GRAPHITE_IMAGE(x;y)'	
SEM(98)	'EXTRACT(x,y)'	

FIGURE 7.2 *Example of a* GPFM *lexical entry from (Spencer, 2005).*

(SYN), and a semantic representation (SEM), as well as a lexemic index (LI) which constitutes a unique identifier (usually represented as an integer) for each individual lexeme. As an example Spencer (2005) cites two homophonous verbs: a first verb DRAW meaning 'make a pictorial image using a graphite medium,' associated with the (arbitrary) lexemic index 87, and a second verb DRAW meaning 'pull, drag, extract,' associated with the (arbitrary) lexemic index 98. The formal representation for each dimension is illustrated in Figure 7.2.

At the MOR(phological level), the representation in Figure 7.2 indicates that both verbs share the same set of stems *draw, drew,* and *drawn* which are associated with specific cells in the verbal paradigm (Past, Past Participle, and, implicitly, all the others; in other words, the default) and the same morphological category (V). This MOR information corresponds to the stem and class information given by Matthews. Spencer's representation however goes beyond specifying lexical information as relevant to inflectional morphology. The dimensions associated with lexical entries also include syntactic and semantic information. In terms of their SYN(tax), the two homophonous verbs thus also share the same argument structure (a subject and a direct object). But at the level of SEM(antics), they differ in their respective meanings.

Lexical Entries in NM (Brown and Hippisley, 2012)

Network Morphology (NM: Brown and Hippisley 2012) shares with GPFM the fact that it is not restricted to inflectional morphology. Work in NM also extends to derivational morphology.[6] NM is typically implemented and represented in the lexical knowledge representation language DATR (Evans and Gazdar, 1989, 1996). As such, the overall architecture of lexical representation in NM is highly inheritance-based—making extensive use of default inheritance.

6 See for example (Hippisley, 1997) for a study of Russian derivational morphology using NM.

The fundamental idea is that language in general, and morphology and the lexicon in particular, can be represented as a default inheritance network. Grammar is conceived as a cascade of defaults which, in some instances, can be replaced by specific exceptions.

Lexemes inherit all the properties of their parent categories. For example, the verb DRAW mentioned above, being a verb, inherits all the basic properties of a verb, unless otherwise stated. An English noun would have the basic properties of the category NOUN and inflect for the singular and plural number. The lexeme SHIRT, as a noun, would inherit that property and have both the forms *shirt* and *shirts* in its paradigm. A noun like TROUSERS however would need to be lexically specified as an exception that does not possess a singular form.

Where primary defining dimensions are concerned, the representation of lexical entries in NM is done in a very similar fashion to GPFM. Overall, lexemes are specified (either directly or by the way of inheritance) through morphological category membership (NOUN or VERB or ADJECTIVE and so on) corresponding to the label MorCat in GPFM. They would also be associated with specific realization rules indicated under mor (specified for a given MorCat in GPFM and PFM), a syntactic category <syn cat> (related to GPFM: SYN), and a semantic category <sem cat> (GPFM: SEM). Stems are often (but not necessarily) represented as a root (<root>) and an association with specific stem selection rules (<stem>). Together they can be seen as the equivalent of the stem inventory under MOR in GPFM. The lexemic identifier in GPFM is comparable to the specific node that each lexical entry represents in the lexical hierarchy.

For example, the Latin verbal lexeme AMO love is represented as in Figure 7.4 by Brown and Hippisley (2012).[7] By default (represented as <> ==), it inherits a large part of its properties from its morphological category (and parent node in the inheritance hierarchy) VERB (<> == VERB). The parent node VERB is represented in Figure 7.3.[8] In this case, the properties Amo inherits from VERB comprise the set of realization rules associated with the active and

---

7 Example from http://networkmorphology.as.uky.edu/sites/default/files/ch5_latin.dtr as of June 2015.

8 About the notation in Figure 7.3: <syn> == "<mor>" stipulates that for verbs, syntactic information typically aligns with whatever is specified for morphological information. Morphological information is structured in the realization of two subparadigms: <mor active> for active forms and <mor passive> for passive forms. The realization rules for those forms are themselves listed by different nodes in the grammar hierarchy. <mor active> == ACT_FORMS:<> and <mor passive> == PASS_FORMS:<> indicate the paths towards those two realization nodes.

```
VERB:
    <syn> == "<mor>"
    <mor active> == ACT_FORMS:<>
    <mor passive> == PASS_FORMS:<>.
```

FIGURE 7.3 *Definiton of the morphological category* VERB *in Network Morphology, Latin example from (Brown and Hippisley, 2012).*

```
Amo:
    <> == VERB
    <gloss> == love
    <root> == am
    <stem> == CONJ_1.
```

FIGURE 7.4 *Lexical entry for the latin verb* AMO *'love' in Network Morphology, example from (Brown and Hippisley, 2012).*

passive forms `<mor active> == ACT_FORMS:<>` and `<mor passive> == PASS_FORMS:<>` and the syntactic properties SYN.[9]

Due to the intrinsic requirements that stem from its computational foundations, NM seems to be the most explicit realizational framework available to date. Computational implementation also offers the benefit of allowing for immediate analytical coherence checking, just as it makes it possible to develop large scale implementations that would, at least in principle, cover all subtle lexical differences in a described language system.

Representing lexical information in the form of general inheritance networks also offers great analytical flexibility in terms of information distribution in addition to generalizability over full systems through carefully designed default cascades. On the other hand, a more rigid representation, as proposed for example in GPFM, will more readily allow us to address questions regarding descriptive primitives and basic notions relevant to the description of morphological systems. The model that we will be describing in the next section is therefore closer in spirit to a GPFM design with a definition of lexical entries based on a small set of descriptive primitives. Being also fully implemented, it

9 In addition to the information mentioned above, the lexical entry Amo also indicates a gloss specification gloss == love, which plays no role in the present paper.

does however share NM's explicitness and potential for large-scale coverage of grammatical descriptions.[10]

PARSLI and the Realizational Lexicon

The general goal in developing the new model described in this section is to offer a tool for approaching WP morphology from a lexical perspective that takes as a sole empirical starting point the paradigmatic properties of lexemes across languages and the dimensions and degrees of variation between them. Analytical concepts such as inflection classes are meant to be definable from a small set of descriptive primitives conditioning variation across paradigms. The resulting model, named PARSLI, PARadigm Shape & Lexicon Interface, is a formal model of inflectional morphology in general and of the interface between the shape of a lexeme's paradigm and the structure of its lexical entry in particular.[11]

PARSLI has been designed to be the lexical counterpart to existing realizational approaches which mainly focus on rule architecture—in other words, on realizational grammar. As indicated above, the goal of realizational approaches is to efficiently account for all (potential) word forms and complete paradigm realization in a given language. Such an approach distinguishes between a REALIZATIONAL GRAMMAR and the REALIZATIONAL LEXICON. The former specifies the realization rules applicable to sets of lexemes in order to generate their full paradigms, as well as other possible types of inflectional information that is shared by sets of lexical items; the latter comprises lexeme-specific information that cannot be represented in terms of the grammar's general patterns.

From a realizational point of view, accounting for inflectional variation can be seen as tracking the number (and proportion) of lexical items that do not comply with the general rule patterns shared by most. Variation amounts to irregularity, that is outliers with deviating paradigm shapes, whether because they are associated with different sets of rules or because they display outright suppletive forms or defective or overabundant paradigms. From this perspective, defining the inflectional lexicon amounts to defining the minimum

10 As can be seen from existing Alexina PARSLI implementations, the potential for large-coverage can be met through the dedicated implementation framework. The best illustration to this point has been the development of DeLex, an Alexina PARSLI based electronic lexical resource for German covering over 63,017 lexemic entries, corresponding to 2.3 million inflected forms (Sagot, 2014).

11 Earlier versions of the framework have been introduced in (Walther, 2011, 2013).

amount of specific lexical information necessary (in addition to the set of regular rules in the grammar) to account for the paradigm realization of any (available) lexical item. As a lexical counterpart to a realizational grammar, PARSLI thus has to account for individual variation and observable types of irregularity. In doing so, it addresses (or readdresses) a series of questions that are central to WP approaches:

1. What relates a specific set of lexemes to their respective paradigms?
2. How can we account for the specific properties of individual lexemes through lexical information and a realizational grammar?
3. Given differences in regularity across a language's paradigms, what lexical (i.e., lexeme specific) information accounts for precisely these irregularities?
4. Can we measure the amount of inflectional irregularity in an inflectional system described within a realizational framework?
5. In what ways do lexemes' inflectional properties cluster?

It is also through these questions that we will be able to offer precise, lexically integrated, formalized, and quantifiable definitions for the fundamental inflectional notions mentioned above.

Lexical Entries: A First Sketch

As PARSLI has been designed to constitute the lexical counterpart to existing realizational approaches, the definition of PARSLI lexical entries naturally presents a certain number of similarities with the definition of traditional realizational lexical entries, as in (Matthews, 1965) or within GPFM (Spencer, 2013). Figure 7.5 illustrates the structure of a (simplified) lexical entry within PARSLI for the English nominal lexeme SHIRT.

If we compare the representation to that of GPFM lexical entries (see Figure 7.2), the whole PARSLI entry only covers a subpart of a GPFM entry, namely the part involved in inflectional morphology within MOR. In other words, a PARSLI lexical entry is an entry of lexemic inflectional properties only—what Fradin and Kerleroux (2003) call an INFLECTEME. An inflecteme is part of a lexeme's definition and corresponds to its paradigmatic properties only. In the GPFM entries in Figure 7.2 above, DRAW_{87} and DRAW_{98} correspond to a unique inflecteme, but two lexemes. The version of PARSLI described in this paper does not represent any syntactic and semantic information or specify an entry's argument structure, as a discussion of these components is beyond the scope of the present paper.

SHIRT	
I-PHON	/ʃəːt/
I-CAT	*noun*
MSF	{ SG, PL }
S-STEM	| (*empty*)
S-FORM	| (*empty*)
I-PAT	| (*see below*)

A PARSLI lexical entry is structured into six defining elements:

(1) I-PHON: an initial phonological input to which realisation rules will be applied for building the individual word forms of the paradigm → inflectional phonological input

(2) I-CAT: a part of speech as relevant to inflection → inflectional category

(3) MSF: a set of morphosyntactic features → morphosyntactic feature bundles

(4) S-STEM: irregular stems that listed instead of built using the rules within the grammar → suppletive stems (empty when regular)

(5) S-FORM: irregular word forms that are listed instead of built using the rules within the grammar → suppletive forms (empty when regular)

(6) I-PAT: a structured set of realisation rules realising the MSF → inflectional pattern

Note: The vertical bars in S-STEM, S-FORM, and I-PAT indicate that the information to their right is meant to be read as an independent object with respect to the overall matrix representation, yet internally structured. S-STEM and S-FORM are overrides of the structured pattern information in I-PAT.

FIGURE 7.5 *Simplified representation of a realisational lexical entry within PARSLI.*

In terms of parallels with existing formalizations of lexical entries, the PARSLI entry in Figure 7.5 must be compared with the formalization of inflectional properties only, such as the MOR part of a GPFM entry. I-CAT appears to play a similar role as that of the MorCat in GPFM, or the parent node in NM. I-PHON combined with S-STEM can be compared to the stem inventory in MOR, the root of stem indications in NM and the stem indices in PFM. In addition, the label at the top left corner of the lexical entry matrix plays the same role as the lexemic identifiers (LI) in GPFM or the lexical ID (LID) in Sign Based

Construction Grammar (SBCG: Boas and Sag 2012), but for inflectional information only. Each ID constitutes a unique label for one particular inflectional lexical entry. The two verbs DRAW$_{87}$ and DRAW$_{98}$ would have a unique inflectional identifier. However, verbs like the auxiliary CAN and the denominal verb CAN meaning 'to put in a can' would have two different identifiers since they do not have the same paradigmatic properties. They could be modeled as in the simplified entries in (5) and (6).

(5) CAN$_1$ (AUX)

I-PHON	/kan/
I-CAT	*verb*
MSF	{ PRS, PST }
S-STEM	(*empty*)
S-FORM	PST: *could*
I-PAT	(*see below*)

(6) CAN$_2$ (DENOM)

I-PHON	/kan/
I-CAT	*verb*
MSF	{ (*standard*) }
S-STEM	(*empty*)
S-FORM	(*empty*)
I-PAT	(*see below*)

At this point, we already see that PARSLI entries offer a representation of lexical information relevant to paradigm realization that is more fine-grained than PFM and GPFM and more stable than NM. This is particularly true of the distinction of suppletive stems (S-STEM) and suppletive forms (S-FORM),[12] as illustrated in (5).

Each of the six defining dimensions within the PARSLI lexical entry has been posited as a locus of potential variation, that is deviation from one or several general patterns stipulated as a larger generalization within the realizational grammar.

While S-STEM and S-FORM, and in a slightly different way MSF, constitute overrides of the general patterns in the grammar—much like local exceptions in NM override the general inheritance structure—the inflectional category

12 See Bonami and Boyé (2002) and Boyé (2006) for a discussion of the difference between the two.

I-CAT and the inflectional pattern I-PAT represent the lexeme-specific anchoring to relevant patterns in the inflectional grammar. The following sections will develop the relationship between lexeme-specific and -general grammatical information, whereby giving an overview of the variety of possible lexemic pattern singularities.

Identifying Descriptive Primitives

Defaults in PARSLI

PARSLI incorporates a default-based representation of regularity similar to that of systematic regular inheritance in NM: regularity in PARSLI corresponds to default inheritance of particular properties associated with a lexeme's inflectional category. Indeed, every lexical item is associated with one particular inflectional category (I-CAT) whose properties it inherits.

Inflectional categories are represented as in (7). A perfectly regular inflectional lexical entry belonging to the I-CAT $\mathscr{C}_i$ in (7) would look like the entry in (8). The indication under I-PAT in the entry for $\mathscr{C}_i$ specifies the set of rules necessary for building the word-forms of regular lexemes of category $\mathscr{C}_i$, thus playing a similar role to PFM (and GPFM) inflection class indications and NM `<mor>` paths. A first notable difference however is that I-PAT specifies rules relevant to stems (here as (Z^s, id)) and to their combination with the right exponence (here as (Z^{exp}, id)), and does not establish a formal difference between stem selection and rules of exponence as commonly done in PFM/GPFM.[13] We will develop the detailed architecture of I-PAT and its relationship to the definition of inflection classes in more detail later on.

(7)
$$\begin{bmatrix} \mathscr{C} & & \\ \text{I-CAT} & \mathscr{C}_i & \\ \text{MSF} & \{ & \sigma_1, \sigma_2, \sigma_3, \ldots \ \} \\ \text{I-PAT} & | & (Z^s, \mathrm{id}), \quad (Z^{exp}, \mathrm{id}) \end{bmatrix}$$

13 The idea of a continuity between stem selection and form realization is also present in Kiparsky's model of Lexical Phonology (Kiparsky, 1982), albeit under otherwise significantly different assumptions regarding the status of this continuity. In particular, Kiparsky's model does not focus on paradigm structure and intra-paradigmatic form co-variance.

(8)

i	
I-PHON	*phon*
I-CAT	$\mathscr{C}_i$
MSF	{ *standard* }
S-STEM	| *(empty)*
S-FORM	| *(empty)*
I-PAT	| *standard*

Feature Inheritance

There are four default relations between an inflectional lexical entry and its inflectional category. The first default concerns the definition of a lexeme's set of morphosyntactic feature bundles that are to be realized in its paradigm. Each inflectional category typically expresses a certain set of morphosyntactic feature bundles (or *morphosyntactic property sets* in PFM). Latin nouns, for example, will be traditionally described as expressing combinations of two number values (SG and PL) and five different case values (NOM, ACC, GEN, DAT and ABL), as illustrated in (9). If a lexeme belongs to a specific category, it will canonically express the same set of features and be marked as *standard* in its lexical entry. In (10), we illustrate this with the inflectional entries for the Latin inflectional category 'noun' and the specific entry for AMĪCUS 'friend' from Table 7.1 page 164.

(9)

NOUN	
I-CAT	*noun*
MSF	{ NOM.SG, VOC.SG, ACC.SG., GEN.SG, DAT.SG, ABL.SG, NOM.PL, VOC.PL, ACC.PL., GEN.PL, DAT.PL, ABL.PL }
I-PAT	| $(Z^s{}_{reg}$, id), $(Z^{exp}{}_{default}$, id)

(10)

AMĪCUS	
I-PHON	amic
I-CAT	*noun*
MSF	{ standard }
S-STEM	| (empty)
S-FORM	| (empty)
I-PAT	| $(Z^s{}_{reg}$, id), $(Z^{exp}{}_{CL\text{-}O}$, id)

A lexical entry which does not conform to the default would specify its own set of morphosyntactic feature bundles (MFSs). MFS deviation can be of three types: a lexeme can have a paradigm expressing more morphosyntactic feature bundles than regular, or less morphosyntactic feature bundles than regular, or it can present morphosyntactic mismatches. We will come back to the third pattern later and for now concentrate on the expression of more or less features than expected. For irregular lexical items in this sense, these deviations will be noted under MFS in their lexical entry. In a canonical approach to regularity, they would be said to display the non-canonical phenomena of OVERDIFFERENTIATION or DEFICIENCY (Walther, 2013).[14] Examples are given in (12) and (13) below for the Slovak noun PRIATEĽ 'friend' and the French verb SEOIR 'be convenient for someone'.

Overdifferentiation

Slovak nouns typically inflect for case (NOM, GEN, DAT, ACC, LOC, and INS) and number (SG and PL), as illustrated by the noun BÁDATEĽ 'researcher' in Table 7.2. Zauner (1973) however indicates that there are a certain number of lexemes which also display an additional vocative form in the singular.[15] Table 7.2 gives an example of this exceptional pattern for the noun PRIATEĽ 'friend'.

TABLE 7.2 *Paradigms for Slovak masculine animate consonental class nouns: regular* BÁDATEĽ *'researcher' and overdifferenciating* PRIATEĽ *'friend' (data from Zauner, 1973)*

	BÁDATEĽ 'researcher'		PRIATEĽ 'friend'	
	SINGULAR	PLURAL	SINGULAR	PLURAL
NOM	*bádateľ*	*bádatelia*	*priateľ*	*priatelia*
GEN	*bádateľa*	*bádateľov*	*priateľa*	*priateľov*
DAT	*bádateľovi*	*bádateľom*	*priateľovi*	*priateľom*
ACC	*bádateľa*	*bádateľov*	*priateľa*	*priateľov*
VOC	–	–	*priateľu*	–
LOC	*bádateľovi*	*bádateľoch*	*priateľovi*	*priateľoch*
INS	*bádateľom*	*bádateľmi*	*priateľom*	*priateľmi*

14 The difference between DEFICIENCY and DEFECTIVENESS is defined and briefly discussed at the end of the section on defectiveness and overabundance below.

15 Zauner (1973) lists ten lexemes with this pattern: SYN 'son' (vocative *synku*), BRAT 'brother' (vocative *brat(k)u*), CHLAPEC 'boy' (vocative *chlapče*), ŠVAGOR 'brother-in-law' (vocative *švagre*), KMOTOR 'godfather' (vocative *kmotre*), CHLAP 'guy' (vocative *chlape*), PÁN 'Mister' (vocative *pane*), MAJSTER 'master' (vocative *majtre*), BOH 'god' (vocative *bože*), ČLOVEK 'individual' (vocative *človece*).

The corresponding lexical entry for PRIATEĽ is given in the example (12). Its set of morphosyntactic feature bundles is irregular and thus explicitly specified in its MSF indication: it indicates the additional VOC.SG thereby overriding the Slovak NOUN inflectional category default represented in the reduced entry for a typical Slovak NOUN_{MSF} in (11).

(11)

NOUN_{MSF}	
MSF	{ NOM.SG, GEN.SG, DAT.SG, ACC.SG, LOC.SG, INS.SG, NOM.PL, GEN.PL, DAT.PL, ACC.PL, LOC.PL, INS.PL }

(12)

PRIATEĽ	
I-PHON	priateľ
I-CAT	*noun*
MSF	{ NOM.SG, GEN.SG, DAT.SG, ACC.SG, VOC.SG, LOC.SG, INS.SG, NOM.PL, GEN.PL, DAT.PL, ACC.PL, LOC.PL, INS.PL }
S-STEM	(*empty*)
S-FORM	(*empty*)
I-PAT	(z^{s}_{c}, id), $(Z^{exp}_{\text{M.ANIM-C}}, \text{id})$ (z^{s}_{c}, id), $(z^{exp}_{\text{VOC}}, \text{id})$

Deficiency

The case where a lexeme's paradigm appears to express less morphosyntactic feature bundles than expected can be illustrated with the French verb SEOIR be 'convenient for someone'. French verbal paradigms usually comprise 47 different cells (see for example the French conjugation dictionary Bescherelle (Arrivé, 1997)). A verb like SEOIR however, whose paradigm is given in Table 7.3, shows a pattern with missing forms. According to the Bescherelle conjugation dictionary, SEOIR's paradigm only comprises thirteen word-forms. Its irregular lexical entry explicitly specifies the morphosyntactic feature bundles for which SEOIR inflects (indicated in (13)) thereby overriding the French VERB_{MSF} inflectional category default.

(13)

$\text{SEOIR}_{\text{MSF}}$	
MSF	{ 3SG.PRS.IND,3PL.PRS.IND, 3SG.PST.IND, 3PL.PST.IND, 3SG.FUT.IND,3PL.FUT.IND, 3SG.PRS.COND, 3PL.PRS.COND, 3SG.PRS.SBJV, 3PL.PRS.SBJV, INF, PTCP }

TABLE 7.3 *The deficient paradigm of French* SEOIR *'be convenient for someone' according to Arrivé (1997)*

	PRS	PST (IPRF)	FUT
3SG.IND	*sied*	*seyait*	*siéra*
3PL.IND	*siéent*	*seyaient*	*siéront*
3SG.COND	*siérait*		
3PL.COND	*siéraient*		
3SG.SBJV	*siée*		
3PL.SBJV	*siéent*		
INF	*seoir*		
PTCP	*séant/seyant*	*sis*	

Pattern Inheritance

Another type of default discernible in the examples we have given so far concerns the structure of the inflectional pattern (I-PAT). Inflectional categories can be specified with a particular inflectional pattern corresponding to sets of realization rules used, by default, for realizing all the specified morphosyntactic feature bundles in their MSF information. Irregularity can be indicated by overriding this default I-PAT with a lexeme specific inflectional pattern (as for example in the entries for Latin AMĪCUS 'friend' in (10) and Slovak PRIATEĽ 'friend' in (12). The next section gives a more detailed description of the content of PARSLI inflectional patterns.

Suppletion as an I-PAT Override

The two remaining defaults in lexical entries pertain to the absence of suppletive forms. By default, all word forms are built through the default realization rules associated with a lexeme's inflectional category's inflectional pattern. In addition to overriding a lexeme's default pattern with a lexeme specific pattern, lexical entries may also specify sets of suppletive stems or suppletive forms. Those are indicated under S-STEM and S-FORM. By default, the S-STEM and S-FORM lists are marked as *empty*, but lexemes with suppletive elements can specify values.

An example of the representation of form suppletion has been given in the simplified lexical entry for the auxiliary CAN, noted CAN_1 in example (5), which explicitly specifies a suppletive form *could* for its past tense. An example of stem suppletion is given for the French verb ALLER 'go' in (14) below. ALLER has four irregular stems that appear in different parts of its paradigm. Here, they are listed with corresponding (arbitrary) stem indices. S-STEM can be seen as a formal representation of what Aronoff (1994) refers to as 'indexed stems' in the description of Latin verbal inflection.

(14)

ALLER		
I-PHON	all	
I-CAT	*verb*	
MSF	{ standard }	
S-STEM	1:	*all-*
	2:	*v-*
	3:	*aill-*
	4:	*i-*
S-FORM	(*empty*)	
I-PAT	(*see below*)	

Realization Zones and (Partial) Form Covariation

As a realizational model, PARSLI represents morphological information relevant to form realization as distributed over a realizational grammar and an inflectional lexicon populated with inflectional lexical entries. Within the realizational grammar, observable regularities primarily concern the association of sets of realization rules with lexical entries. In particular, paradigmatic regularities will reveal that certain realization rules systematically appear together in sets of rules used for the realization of a given paradigm (or subparadigm). For example, in the case of Latin noun inflection, we observe that the rules suffixing *-a, -am, -ae, -ās* and *-ārum* to the noun stem are always used together by a subset of nominal lexemes, namely those traditionally described as belonging to the class of a-stem nouns. Similarly *-us -e, -um, -ī, -ō, -ōs* and *-ōrum* always appear jointly for realizing o-stem paradigms. The rules suffixing *-īs* in the dative and ablative plural however, are shared by both subclasses of Latin nouns. Examples of a class-a and a class-o noun are given in Table 7.4.

Suppose we wanted to account for a (simplified) system of Latin noun inflection which comprises only two types of nouns: o-stem nouns and a-stem nouns.[16] We could first define the two separate sets of realization rules that systematically cluster together depending on the paradigm type (o-stem or a-stem paradigms). They would realize a subpart of the nominal paradigms, namely those associated with the following MSF: {NOM.SG, VOC.SG, ACC.SG,

16 The actual Latin noun inflection system is of course more complex, but this simplified version provides a more transparent illustration of how realization zones can be defined.

TABLE 7.4 *Latin paradigms for o-stem and a-stem class nouns:* AMĪCUS *'(male) friend' and* AMĪCA *'(female) friend'*

	SG	PL	SG	PL
NOM	amīcus	amīcī	amīca	amīcae
VOC	amīce	amīcī	amīca	amīcae
ACC	amīcum	amīcōs	amīcam	amīcās
GEN	amīcī	amīcōrum	amīcāe	amīcārum
DAT	amīcō	amīcīs	amīcae	amīcīs
ABL	amīcō	amīcīs	amīcā	amīcīs

GEN.SG, DAT.SG, ABL.SG, NOM.PL, VOC.PL, ACC.PL, GEN.PL}. In addition to these initial rule sets, we would need one supplementary set of rules that could be combined with either of the former two in order to realize the remaining morphosyntactic feature bundles in the nominal paradigms, namely the dative and ablative plural.

This idea of systematic clustering of certain rules is the origin of the notion of REALIZATION ZONE in PARSLI. Realization zones are generalizations over partial paradigms that express the covariation of forms and, when representing systems in realizational terms, of rules. Each lexeme is associated with a certain number of realization zones, comprising realization rules used for their paradigm realization. The realization zones corresponding to our simplified Latin noun inflection are listed in Table 7.5. A noun like the o-stem noun AMĪCUS '(male) friend' would be associated with the zones z_{O}^{exp} and z_{ALL}^{exp}, a noun like the a-stem noun AMĪCA '(female) friend' would be associated with the zones z_{A}^{exp} and z_{ALL}^{exp}. This has the advantage of not having to stipulate two separate realization rules for the dative or ablative plural, one in the a-class and one in the o-class, thus *explicitly highlighting that the two paradigms share part of their form realization properties*. With this form of representation, the same rule set (for {DAT.PL, ABL.PL}) need only be posited once and can then be combined with other rules from the two other zones to realize the two types of paradigms.

The lexical entries for AMĪCUS and AMĪCA would contain the inflectional patterns (I-PAT) in (15) and (16) below. In these I-PATs, (z_{reg}^{s}, id) stands for the realization rules responsible for stem realization, in this particular case the regular (invariant) stem inherited from the I-PAT of the nominal lexical entry illustrated in (9) above.

TABLE 7.5 *Realisation zones for a simplified Latin noun inflection system, restricted to rules of exponence*

REALISATION ZONE	EXPRESSED MORPHOSYNTACTIC FEATURES
z_{O}^{exp}	{NOM.SG, VOC.SG, ACC.SG., GEN.SG, DAT.SG, ABL.SG, NOM.PL, VOC.PL, ACC.PL., GEN.PL}
z_{A}^{exp}	{NOM.SG, VOC.SG, ACC.SG., GEN.SG, DAT.SG, ABL.SG, NOM.PL, VOC.PL, ACC.PL., GEN.PL}
z_{ALL}^{exp}	{DAT.PL, ABL.PL}

While (15) and (16) form the inflectional patterns of AMĪCUS and AMĪCA, each line in that pattern, that is each combination of a stem realization zone[17] and a realization zone of exponence, will be called a SUBPATTERN.

(15)
$$\left[\begin{array}{l|ll} \text{AMĪCUS}_{\text{I-PAT}} & (Z^{s}{}_{reg}, \text{id}), & (z_{O}^{exp}, \text{id}) \\ \text{I-PAT} & (Z^{s}{}_{reg}, \text{id}), & (z_{ALL}^{exp}, \text{id}) \end{array}\right]$$

(16)
$$\left[\begin{array}{l|ll} \text{AMĪCA}_{\text{I-PAT}} & (Z^{s}{}_{reg}, \text{id}), & (z_{A}^{exp}, \text{id}) \\ \text{I-PAT} & (Z^{s}{}_{reg}, \text{id}), & (z_{ALL}^{exp}, \text{id}) \end{array}\right]$$

Inflectional Classes as a Secondary Generalization

From what has been said so far, we see that realization zones represent descriptive primitives in a PARSLI representation. It is the zones, as a locus of systematic form covariation, and not the inflection classes, that are primarily associated with a particular lexical entry within its inflectional pattern I-PAT. Inflection classes, such as the a-stem and the o-stem nominal inflection classes here, are secondary generalizations about frequently (but not systematically) clustering realization zones. In other words, they correspond to frequent covariation of sets of forms (or rules). Contrary to PFM/GPFM, NM, or the lexical entry definition by Matthews (1965), inflectional classes do not constitute descriptive primitives. They can however be derived from the frequency of zone clustering.

17 Or (in this case) class, as defined in the next section below.

Once a particular secondary generalization, i.e., an inflection class, has been found to be useful as a means for capturing recurrent patterns of covariation in a specific language, a lexeme can be associated in its I-PAT with an inflection class directly. In our Latin examples, the clustering of z_{O}^{exp} and z_{ALL}^{exp} is a frequent one. We can thus posit the o-stem inflection class $Z_{\mathrm{CL\text{-}O}}^{exp}$ defined as in (17) and stipulate it directly in the I-PAT of o-stem nouns. The lexical entry for AMĪCUS in (10) does exactly that.

(17) $Z_{\mathrm{CL\text{-}O}}^{exp} = z_{\mathrm{O}}^{exp} \cup z_{\mathrm{ALL}}^{exp}$

The definition of inflection classes is seen as a property of inflectional systems. It is meant to capture significant pattern frequency. As such, it does not constitute lexeme-specific information and is therefore specified within the inflectional grammar. Once regular zone clustering has been specified by positing inflectional classes, lexical entries can make reference to those classes instead of the zones they are built from, which results in more synthetic (by way of higher-level) representations of concerned lexical entries. Evaluating the significance of zone clustering can be achieved through a dedicated descriptive economy measure introduced by Sagot and Walther (2011) and described in the last section of this paper.

Heteroclite Paradigms

Defining inflection classes as a secondary generalization from realization zones also allows for a straightforward representation of the lexical properties of so called HETEROCLITE LEXEMES, i.e., lexemes whose paradigms pattern with more than one inflection class.

As an example, let us look at the inflection of some Slovak animal nouns. Traditionally, Slovak nouns are described as inflecting according to different inflection classes, including one for masculine animate nouns such as CHLAP 'guy' and masculine inanimate nouns such as DUB 'oak'. The paradigms of these nouns are shown in Table 7.6. Their entries are mainly standard, meaning they inherit their MSF from their inflectional category NOUN and do not possess any suppletive forms or stems. Their inflectional patterns, illustrated in (18) and (19), make reference to the masculine animate inflection class Z_{ANIM}^{exp} and to the masculine inanimate inflection class Z_{INAN}^{exp}.

(18) $\begin{bmatrix} \mathrm{CHLAP}_{\text{I-PAT}} & \\ \text{I-PAT} & (Z_{C\ reg}^{S}, \mathrm{id}),\ (Z_{\mathrm{ANIM}}^{exp}, \mathrm{id}) \end{bmatrix}$

TABLE 7.6 *Heteroclisis in Slovak animal nouns* (*Zauner, 1973*)

	Z^{exp}_{ANIM}: MASC. ANIM. CHLAP 'man, guy'		Z^{exp}_{INAN}: MASC. INANIM. DUB 'oak tree'		MASC. HETEROCLITE OROL 'eagle'	
	$z^{exp}_{\text{ANIM,sg}}$: SG	$z^{exp}_{\text{ANIM,pl}}$: PL	$z^{exp}_{\text{INAN,sg}}$: SG	$z^{exp}_{\text{INAN,pl}}$: PL	$z^{exp}_{\text{ANIM,sg}}$: SG	$z^{exp}_{\text{INAN,pl}}$: PL
NOM	*chlap*	*chlap-i*	*dub*	*dub-y*	*orol*	*orl-y*
GEN	*chlap-a*	*chlap-ov*	*dub-a*	*dub-ov*	*orl-a*	*orl-ov*
DAT	*chlap-***ovi**	*chlap-om*	*dub-u*	*dub-om*	*orl-***ovi**	*orl-om*
ACC	*chlap-a*	*chlap-ov*	*dub*	*dub-y*	*orl-a*	*orl-y*
LOC	*chlap-***ovi**	*chlap-och*	*dub-e*	*dub-och*	*orl-***ovi**	*orl-och*
INS	*chlap-om*	*chlap-mi*	*dub-om*	*dub-mi*	*orl-om*	*orl-ami*

(19)

DUB$_{\text{I-PAT}}$	
I-PAT	($Z^{S}_{C\ reg}$, id), (Z^{exp}_{INAN}, id)

There is however a small number of masculine animal nouns that do not follow either of these two patterns, but rather display a mixed, i.e., heteroclite paradigm. Nouns like OROL 'eagle' (see Table 7.6) inflect like animate nouns in the singular and like inanimate nouns in the plural. In PARSLI terms, their paradigms are built from two zones, a singular zone $z^{exp}_{\text{ANIM,sg}}$ typically part of inflection class Z^{exp}_{ANIM}, and a plural zone $z^{exp}_{\text{INAN,pl}}$ typically part of inflection class Z^{exp}_{INAN}. The result is a lexeme with a heteroclite paradigm whose lexical entry can be readily represented as in (20).

(20)

OROL	
I-PHON	orol
I-CAT	*noun*
MSF	{ *standard* }
S-STEM	(*empty*)
S-FORM	(*empty*)
I-PAT	(Z^{S}_{reg}, id), ($z^{exp}_{\text{ANIM, SG}}$, id) ($Z^{S}_{\text{reg}}$, id), ($z^{exp}_{\text{INAN, PL}}$, id)

Naturally, defining heteroclisis can only be done *after* defining inflectional classes, as it constitutes a violation of those secondary generalizations.

Segmentation, Overlapping Contrasts, and Realization Levels

PARSLI accommodates different types of realization zones, such as realization zones with rules for building stems or realization zones with rules of exponence, without imposing any operational difference between those zones.[18] Realization zones can thus be seen as an extension of the notion of PARTITION CLASSES introduced by Pirrelli and Battista (2000) and the notion of STEM SPACE by Bonami and Boyé (2003) to the domain of exponence. The original notions corresponded to the partition of inflectional paradigms into subparadigms using a common stem, in order to account for patterns of stem allomorphy in (Romance) languages. Latin verbs would typically be associated with three different stems, as schematically illustrated for intransitive verbs in the top-layer of Figure 7.6, which corresponds to three distinct stem forms associated with separate sets of morphosyntactic feature bundles summarized in Table 7.7.

In fact, those different types of realization zones simply constitute a way to highlight zones of partial contrast between full word forms. Realization zones thus correspond to DIMENSIONS OF VARIATION. In the Latin example from the top-layer in Figure 7.6, the highlighted contrast is that of the stem allomorphy. But comparing across Latin lexemes, one will also find zones of contrast in the expression of exponence (bottom-layer) or even in the nature of the so called theme vowel (middle-layer). Figure 7.6 shows how the distribution of those contrasts varies from one layer, i.e., contrast level, to the next. For example, Latin verbs are traditionally grouped into classes according to their particular theme vowel (a, o, i or none), resulting in four regular inflection classes and one mixed class.

Tables 7.8 A and B represent partial paradigms for i-class verb AUDIO 'hear' and the athematic verb DUCO 'lead'.

TABLE 7.7 *Latin stem spaces in an intransitive paradigm*

STEM1	all finite imperfective and imperative forms, the present infinitive, present participle and the gerund
STEM2	all finite perfective forms and the past infinitive
STEM3	the future infinitive and participle, and the supine

18 The labelling into stem, theme, and exponence zones indicated hereafter is a pure matter of terminological convenience and could just as well be discarded.

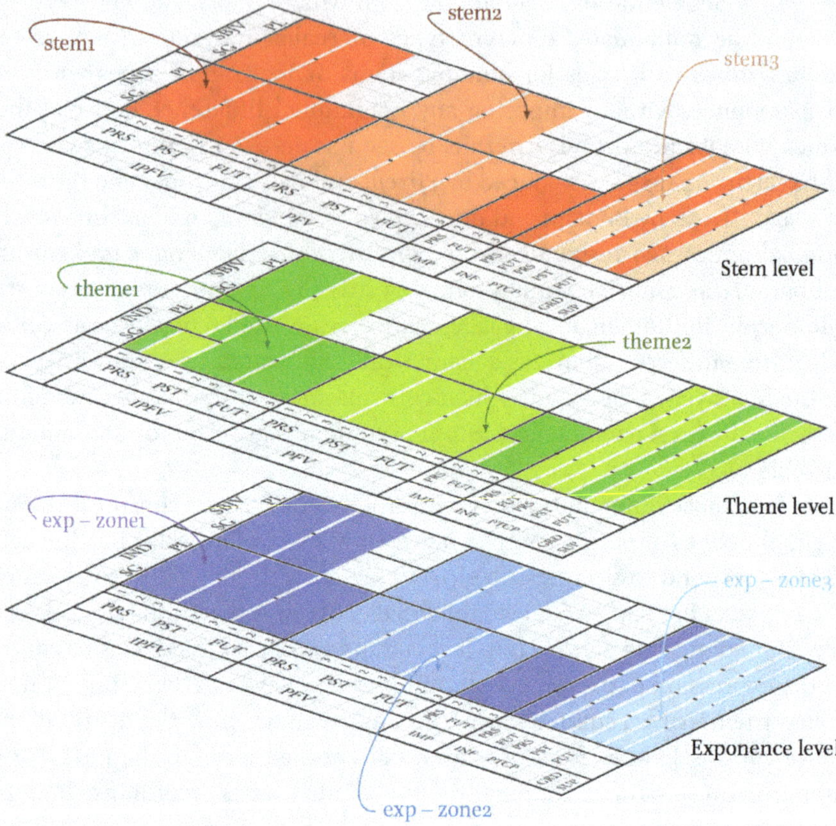

FIGURE 7.6 *A layered representation of contrasts.*

As shown in the partial paradigm of ACCIPIO 'accept' in Table 7.8 C, the mixed class displays the same theme-vowel pattern as the i-class in part of its paradigm and the same pattern as the athematic class in the other part of its paradigm.

In order to represent this fact, we can thus say that standard verbs are associated in their I-PAT with specific theme classes built out of two theme zones each, among which Z_i^{th} for i-class verbs like AUDIO hear and Z_{none}^{th} for athematic verbs like DUCO lead. Mixed verbs are associated with one theme zone from each Z_i^{th} and Z_{none}^{th}, namely $z_{i,1}^{th}$ and $z_{none,2}^{th}$, thus being heteroclite at this level of contrast (or realization). The inflectional patterns for AUDIO, DUCO, and ACCIPIO are indicated in examples (21) to (23).

TABLE 7.8 *Subjunctive imperfective forms of three latin verbs: AUDIŌ 'hear', DUCŌ 'lead', and ACCIPIŌ 'accept'*

I-STEM: AUDIO 'hear'			
		SG	PL
PRS	1	audiam	audiāmus
	2	audiās	audiātis
	3	audiat	audiant
PST	1	audīrem	audīrēmus
	2	audīrēs	audīrētis
	3	audīret	audīrent

A: Subjunctive imperfective forms of latin AUDIŌ 'hear'

ATHEMATIC: DUCO 'lead'			
		SG	PL
PRS	1	ducam	ducāmus
	2	ducās	ducātis
	3	ducat	ducant
PST	1	ducerem	ducerēmus
	2	ducerēs	ducerētis
	3	duceret	ducerent

B: Subjunctive imperfective forms of latin DUCŌ 'lead'

MIXED: ACCIPIO 'accept'			
		SG	PL
PRS	1	accipiam	accipiāmus
	2	accipiās	accipiātis
	3	accipiat	accipiant
PST	1	acciperem	acciperēmus
	2	acciperēs	acciperētis
	3	acciperet	acciperent

C: Subjunctive imperfective forms of latin ACCIPIŌ 'accept'

(21) $\left[\begin{array}{l|lll} \text{AUDIO}_{\text{I-PAT}} & & & \\ \text{I-PAT} & (Z^s_1, \text{id}), & (Z^{th}_i, \text{id}), & (Z^{exp}_B, \text{id}) \end{array}\right]$

(22) $\left[\begin{array}{l|lll} \text{DUCO}_{\text{I-PAT}} & & & \\ \text{I-PAT} & (Z^s_2, \text{id}), & (Z^{th}_{none}, \text{id}), & (Z^{exp}_B, \text{id}) \end{array}\right]$

(22) $\left[\begin{array}{l|lll} \text{ACCIPIO}_{\text{I-PAT}} & (Z^s_3, \text{id}), & (z^{th}_{i,1}, \text{id}), & (Z^{exp}_B, \text{id}) \\ \text{I-PAT} & (Z^s_3, \text{id}), & (z^{th}_{none,2}, \text{id}), & (Z^{exp}_B, \text{id}) \end{array}\right]$

The combination of these zones and their layering creates a lexeme's specific paradigmatic properties (see Figure 7.6). It is represented by the structure of the inflectional patterns (I-PAT) in the inflectional entries. Such contrasts need not be based on form segmentation as for this Latin examples. For other types of data, it might make sense to devote a specific layer to suprasegmental contrasts.[19]

Morphosyntactic Mismatches ("Deponency")

In addition to the paradigmatic peculiarities described above, there is one case that has not been discussed so far, namely that of MORPHOSYNTACTIC MISMATCHES, also commonly known as DEPONENCY (Baerman, 2007b). An example presented by Baerman (2006) involves Serbo-Croatian nouns,[20] which inflect along several different inflection classes, among which a class for neuter nouns with an et-stem Z^{exp}_{N-et} like DUGME 'button', a class for masculine nouns Z^{exp}_M like JELEN 'stag deer', and a feminine a-stem class Z^{exp}_{F-a} as for ŽENA 'woman'. Their paradigms are illustrated in Table 7.9.

Moreover, some (less numerous) irregular nouns such as DETE 'child' or BRAT 'brother' are heteroclite. As can be seen from Table 7.9, their paradigms are built using realization rules from zones belonging to two different inflection classes: the singular zone $z^{exp}_{N-et,sg}$ from class Z^{exp}_{N-et} and the singular $z^{exp}_{F-a,sg}$ zone from class Z^{exp}_{F-a} for DETE, and the singular zone $z^{exp}_{M,sg}$ from class Z^{exp}_M and the singular $z^{exp}_{F-a,sg}$ zone from class Z^{exp}_{F-a} for BRAT. But in addition to being simply heteroclite, those paradigms also show a form of morphosyntactic mis-

19 Such a treatment might be appropriate for Russian stress patterns. The layering could then be compared to the use of parallel hierarchies in the NM account of Russian noun inflection proposed by Brown et al. (1996).

20 The name used for referring to the language reflects the fact that data, registered in the Deponency Database in Surrey (Baerman, 2006), had been collected before the breakup of former Yugoslavia.

TABLE 7.9 *Morphosyntactic mismatch in Serbo-Croatian* (Baerman, 2006)

	$Z^{exp}_{N-et,sg}$: NEUTER –ET STEM DUGME 'button'		Z^{exp}_{M}: MASC JELEN 'stag, deer'		Z^{exp}_{F-a}: FEM –A STEM ŽENA 'woman'	
	$z^{exp}_{N-et,sg}$: SG	$z^{exp}_{N-et,pl}$: PL	$z^{exp}_{M,sg}$: SG	$z^{exp}_{M,pl}$: PL	$z^{exp}_{F-a,sg}$: SG	$z^{exp}_{F-a,pl}$: PL
NOM	*dugme*	*dugm-et-a*	*jelen*	*jelen-i*	*žen-a*	*žen-e*
ACC	*dugme*	*dugm-et-a*	*jelen-a*	*jelen-e*	*žen-u*	*žen-e*
GEN	*dugm-et-a*	*dugm-et-a*	*jelen-a*	*jelen-a*	*žen-e*	*žen-a*
DAT	*dugm-et-u*	*dugm-et-ima*	*jelen-u*	*jelen-ima*	*žen-i*	*žen-ama*
LOC	*dugm-et-u*	*dugm-et-ima*	*jelen-u*	*jelen-ima*	*žen-i*	*žen-ama*
INS	*dugm-et-om*	*dugm-et-ima*	*jelen-om*	*jelen-ima*	*žen-om*	*žen-ama*

	NEUTER –ET STEM/ –A STEM DETE 'child'		MASC / –A STEM BRAT 'brother'	
	$z^{exp}_{N-et,sg}$: SG	$z^{exp}_{F-a,sg}$: PL	$z^{exp}_{M,sg}$: SG	$z^{exp}_{F-a,sg}$: PL
NOM	*dete*	*dec-a*	*brat*	*brać-a*
ACC	*dete*	*dec-u*	*brat-a*	*brać-u*
GEN	*det-et-a*	*dec-e*	*brat-a*	*brać-e*
DAT	*det-et-u*	*dec-i*	*brat-u*	*brać-i*
LOC	*det-et-u*	*dec-i*	*brat-u*	*brać-i*
INS	*det-et-om*	*dec-om*	*brat-om*	*brać-om*

match. The realization rules from realization zone $z^{exp}_{F-a,sg}$ used for building their plural forms are rules which are typically associated with singular forms in the feminine a-stem class Z^{exp}_{F-a}.

In order to account for that lexeme-specific mismatch in the use of otherwise defined realization rules, PARSLI identifies the feature mismatch with a lexeme specific TRANSFER RULE associated with each specified realization zone (or class). In the default case (such as those seen in all examples above) this transfer rule is simply the identity function, noted id. The default case is re-illustrated here for the noun ŽENA in (24).

(24)
$$\left[\begin{array}{l|l} \text{ŽENA}_{\text{I-PAT}} & \\ \text{I-PAT} & (Z^{s}_{reg}, \text{id}), \quad (Z^{exp}_{\text{F-A}}, \text{id}) \end{array}\right]$$

Transfer rules are comparable to paradigm functions in PFM. The default identity function id corresponds to the default rule of paradigm linkage where a morphosyntactic property set σ in a content paradigm cell <*L*,σ> is 'by default' identical to the morphosyntactic property set τ in its form paradigm <*s*,τ>. In NM terms, this corresponds to default identity between syntactic `<syn>` and morphological `<mor>` information.

In cases of mismatches such as in the paradigms of DETE and BRAT, the transfer rule will be a specific one, namely here the rule t_{NB} defined as in (25).

(25) $t_{\text{NB}}(\text{SG})=\text{PL}$

t_{NB} indicates that the expressed value of the rule realising a singular will be plural if associated with a particular lexeme. In this, it functions just as non default rules of Paradigm Linkage in PFM (and GPFM) or exceptions in the paths between the morphological information under `<mor>` and the syntactic information under `<syn>` in NM. As the transfer rule is available for all lexical entries presenting this type of mismatch, its content is specified as a type of pattern within the grammar. Lexical entries can make reference to this rule within their inflectional patterns as in (26) for DETE and (27) for BRAT.

(26)
$$\begin{bmatrix} \text{DETE}_{\text{I-PAT}} & \Big| & (Z^{s}_{\text{reg}}, \text{id}), & (z^{exp}_{\text{N-ET, SG}}, \text{id}) \\ \text{I-PAT} & \Big| & (Z^{s}_{\text{reg}}, \text{id}), & (z^{exp}_{\text{F-A, SG}}, t_{\text{NB}}) \end{bmatrix}$$

(27)
$$\begin{bmatrix} \text{BRAT}_{\text{I-PAT}} & \Big| & (Z^{s}_{\text{reg}}, \text{id}), & (z^{exp}_{\text{M, SG}}, \text{id}) \\ \text{I-PAT} & \Big| & (Z^{s}_{\text{reg}}, \text{id}), & (z^{exp}_{\text{F-A, SG}}, t_{\text{NB}}) \end{bmatrix}$$

Missing or Crowded Cells: Defectiveness and Overabundance

Overabundance

Some inflectional paradigms are sometimes described as having more than one form realizing one morphosyntactic feature bundle. This phenomenon has been labeled OVERABUNDANCE by Thornton (2011). Table 7.10 gives an example from the French verb ASSEOIR 'sit/seat'. According to the description in the Bescherelle (Arrivé, 1997), this pattern of having two possible forms realizing a single morphosyntactic feature bundle applies to about half of the verb's paradigm.[21]

21 Thornton (2011) indicates that true (*canonical* in the sense of Corbett 2007) overabundance concerns cases where multiple form realizations are in true competition without a context, speaker, or additional feature favoring one over the other. Although the account of ASSEOIR given in the Bescherelle corresponds to the currently accepted one, a detailed corpus study would probably still be beneficial in order to evaluate the true

TABLE 7.10 *Overabundance in (a subpart of) the paradigm of* ASSEOIR *'sit/seat'*

IND.PRS	SG	PL
1	*assois* *assieds*	*assoyons* *asseyons*
2	*assois* *assieds*	*assoyez* *asseyez*
3	*assoit* *assied*	*assoient* *asseyent*

As can be seen from Table 7.10, each dual realization of the relevant morphosyntactic feature bundles can easily be accounted for through an association of a regular inflectional ending with two different stems. A clear example is the association of the ending *-ons* with each of the stems *assey* and *assoy*. A way of representing this in a PARSLI inflectional pattern is simply to posit that ASSEOIR's I-PAT comprises supernumerary stem realization zones that are all associated with the expected inflection class for ASSEOIR in their respective subpatterns. This representation is illustrated in (28).

(28)
$$\left[\begin{array}{lll} \text{ASSEOIR}_{\text{I-PAT}} & (z^{s}_{ied}, \text{id}), & (Z^{exp}_{\text{ASSEOIR}}, \text{id}) \\ & (z^{s}_{oi}, \text{id}), & (Z^{exp}_{\text{ASSEOIR}}, \text{id}) \\ \text{I-PAT} & (z^{s}_{ey}, \text{id}), & (Z^{exp}_{\text{ASSEOIR}}, \text{id}) \\ & (z^{s}_{oy}, \text{id}), & (Z^{exp}_{\text{ASSEOIR}}, \text{id}) \\ & (z^{s}_{i}, \text{id}), & (Z^{exp}_{\text{ASSEOIR}}, \text{id}) \end{array}\right]$$

Defectiveness

Other paradigms are described as suffering from missing forms. One way to account for missing forms has been indicated above for cases of deficiency, i.e.,

nature of form competition for ASSEOIR. As a means for illustrating PARSLI representations of overabundance, we will however assume the broadly accepted view, according to which the dual realizations in the paradigm of ASSEOIR are indeed in true competition for (at least some not so rare) contemporary speakers of French. The representation given here is how we would model such a phenomenon as a paradigmatic peculiarity within PARSLI.

cases where a lexical entry does not express all the morphosyntactic feature bundles canonically associated with its inflectional category. However, given what we have just shown for the cases of overabundance, another scenario is also plausible. Just as overabundant lexemes are associated with more than one realization zone for a particular set of morphosyntactic feature bundles, so called DEFECTIVE lexemes are missing a zone for a particular set of morphosyntactic feature bundles.

An example can be drawn from the paradigm of the French verb TRAIRE 'milk'. According to the Bescherelle (Arrivé, 1997), its paradigm is lacking all forms for the simple past indicative and imperfect subjunctive forms.[22] While the endings for these forms are regular across all other French paradigms, the corresponding stems vary. In the case of TRAIRE, no rule defining stem realization can be identified. We may thus say that TRAIRE lacks an association with a stem level realization zone for the morphosyntactic feature bundles comprising {IND.SPST, SBJV.IPRF}. The inflectional pattern for TRAIRE representing this lack of association is given in (29).

(29)
$$\left[\begin{array}{l|ll} \text{TRAIRE}_{\text{I-PAT}} & (z^{s}_{ai}, \text{id}), & (Z^{exp}_{\text{TRAIRE}}, \text{id}) \\ \text{I-PAT} & (z^{s}_{ay}, \text{id}), & (Z^{exp}_{\text{TRAIRE}}, \text{id}) \\ & \text{—} & (Z^{exp}_{\text{SEOIRE}}, \text{id}) \end{array}\right]$$

An obvious question at this point would be how we differentiate between cases of DEFICIENCY and cases of DEFECTIVENESS. An obvious first answer to that question would of course be that it is a question of personal analytical choices leading to focus either on the (lack of) expressed morphosyntactic feature bundles or on the (lack of) existence of realization rules for the paradigms' missing cells.

In this case, compared to that of French SEOIR 'be convenient for someone', one might however point out the following: while it is easy to coin the missing forms of SEOIR whose existing forms are built entirely along the same pattern as those of (non-defective) ASSEOIR 'sit/seat',[23] it is more difficult to decide for TRAIRE whether the first person singular of the indicative simple past should be *je *tris* along the pattern used by FAIRE 'do', *je *trus* as it would be for PLAIRE 'please' or *je *trayai* as for regular first group verbs (which however have a different infinitive form than TRAIRE). This behavioral difference

22 Despite the rather broad acceptance of this account of TRAIRE across current morphological studies of French, it would still be useful to validate this consensus view by verifying the absence of forms of TRAIRE in large corpora. Meanwhile, this section presents the way defectiveness can be represented in PARSLI.

23 However without the additional overabundant forms described above.

could be a criterion for differentiating between two causes for incomplete paradigms: defectiveness for cases where missing cells are not readily mendable (i.e., where the appropriate realization rule seems difficult or even impossible to ascertain), and deficiency for cases where coining the forms from the missing subparadigms would be easy but the forms themselves are rarely attested.[24] Justification of one classification or the other in particular cases could appeal to psycholinguistic studies that probe speaker behavior and intuitions.

Analytical Choices and Writing a Grammar with PARSLI

The previous section has given an overview of the primitives defining PARSLI lexical entries and their relation to the definition of secondary generalizations such as inflection classes or levels of contrast, as well as deviations from those primitives and secondary generalizations such as stem or form suppletion, heteroclite paradigms, or morphosyntactic mismatches. Yet it is clear that the definition of those deviations ultimately depends on analytical choices we make while defining a particular grammar.

Let us for example go back to the Slovak heteroclite paradigm example: we have presented the data as a case where there are four realization zones which constitute two inflection classes associated with two sets of nouns: the animate and the inanimate masculine nouns. An additional set of (animal) nouns has been described as heteroclite, building its paradigms using one zone from each class. But with the same four realization zones, we could just as easily have defined three inflection classes defined as in (30) to (32). In this case, two of the four zones would be participating in the definition of two inflection classes each—just as proposed for the realization zone expressing {DAT.PL, ABL.PL} in our simplified Latin example in (17). The question we face is thus how to choose between those two options.

(30) $Z^{exp}_{\text{ANIM}} = Z^{exp}_{\text{ANIM, SG}} \cup Z^{exp}_{\text{ANIM, PL}}$

(31) $Z^{exp}_{\text{INAN}} = Z^{exp}_{\text{INAN, SG}} \cup Z^{exp}_{\text{INAN, PL}}$

(32) $Z^{exp}_{\text{ANIMAL}} = Z^{exp}_{\text{ANIMAL, SG}} \cup Z^{exp}_{\text{ANIMAL, PL}}$

24 Of course, no corpus study, no matter how extensive, will ever offer an exhaustive inventory of all word forms. What is however interesting in the case of TRAIRE is that according to the accepted accounts, there are entire subparadigms missing which are featurally coherent, including forms realizing morphosyntactic feature bundles that are not intrinsically rarely realized, as for example the forms of the indicative simple past.

Similar decisions have to be made when accounting for possible cases of suppletion or morphosyntactic mismatches. For example: how do we want to account for English irregular verbs? As verbs with a set of suppletive stems as in (33) or as verbs sharing a specific stem realization zone for building their past form and past participle as in (34). This specific stem realization zone (noted Z^{s}_{apo} in (34)) would be based on an apophonic i/a/u alternation, also shared by verbs like DRINK or SWIM, and thus differ from the main regular pattern.

The first option leads to more complicated lexical entries, whereas the second requires defining additional realization zones with specific realization rules within the grammar.

(33) SING: suppletion

SING		
I-PHON	/siŋ/	
I-CAT	*verb*	
MSF	{ standard }	
S-STEM	(*empty*)	
S-FORM	PST:	*sang*
	PST.PTCP:	*sung*
I-PAT	(Z^{s}, id),	(Z^{exp}, id)

(34) SING: (MORE) REGULAR

SING		
I-PHON	/siŋ/	
I-CAT	*verb*	
MSF	{ standard }	
S-STEM	(*empty*)	
S-FORM	(*empty*)	
I-PAT	(Z^{s}_{apo}, id),	(Z^{exp}, id)

Analyzing the set of Serbo-Croatian nouns mentioned above also comes down to making definite analytical choices. Instead of positing the transfer rule in (25) within the grammar and invoking it within inflectional patterns for DETE 'child' in (26) and BRAT 'brother' in (27), we could also have stipulated a different realization zone within the grammar which would realize plural but would happen to use the same formal markings as another zone realizing singular morphosyntactic feature bundles.

Let us assume an analysis of English irregular verbs with multiple stem realization zones as suggested through the lexical entry in (34). We have identified a class of verbs which form their past form and past participle through apophony, such as the verbs SING (*sing – sang – sung*), DRINK (*drink – drank – drunk*), or SWIM (*swim – swam – swum*) among others. A verb like CLING (*cling – clung – clung*) could thus either be treated as a verb inflecting according to another independent stem class, or as a verb of the same class as SING, CLING, and SWIM, but with a suppletive past form, or as a verb with a morphosyntactic mismatch that makes it use the past participle form for expressing the past tense. In this latter case, it would require stipulating a transfer rule as in (35), while the inflectional pattern for CLING in an analysis involving such a morphological mismatch would be as in (36).

(35) $t_{\text{IRR-PST}}(\text{PST}) = \text{PST.PTCP}$

(36) CLING: in analysis involving a morphosyntactic mismatch

$$\left[\begin{array}{l|l} \text{CLING}_{\text{I-PAT}} & \\ \text{I-PAT} & (Z^{s}_{apo}, t_{\text{IRR-PST}}),\quad (z^{exp}, \text{id}) \end{array}\right]$$

In order to choose between these options, we need an evaluable criterion. In a realizational description of inflection, the choice should be coherent with the overall goal of realizational approaches: to give an *efficient* account of the paradigmatic properties of inflectional systems in a way highlights all possible generalizations that can be made about the system.

Thus, given multiple possible accounts, the model that would account for all word forms of a system and all paradigmatic interactions between them in the most efficient way, not missing out on any possible generalizations, would be considered the best choice. Ideally, we would have an evaluation metric of the most economic, that is the shortest description of the system. Suppose the analytical choice were to posit an additional inflection class or not. The choice would have to be the one which minimizes the overall length of the description. We would have to evaluate whether we would obtain a shorter overall description by positing an additional inflection class that multiple lexical items could refer to directly (instead of listing all the realization zones), knowing that defining this inflection class within the grammar lengthens the grammatical description. In other words, we would need to find the most economic equilibrium in the distribution of information between the grammar and the lexical entries, i.e., the grammar and the lexicon.

This idea is not new and has been explored in various ways since the enunciation of the concept of EXPLANATORY ADEQUACY proposed by Chomsky (1965) as a means for ranking competing (formally described) accounts of a given grammatical system.

> The major problem in constructing an evaluation measure for grammars is that of *determining which generalizations are significant ones*; an evaluation measure must be selected in order to favor these.
>
> [...]
>
> The *obvious numerical measure* to be applied to a grammar is length, in terms of number of symbols. But if this is to be a meaningful measure, *it is necessary to restrict the form of rules* in such a way that significant considerations of complexity and generality are converted into considerations of length, *so that real generalizations shorten the grammar and spurious ones do not.*
>
> CHOMSKY (1965, 42) [EMPHASIS ADDED GW]

While most past attempts may have been perceived to have resulted in simplistic and predictably unsuccessful symbol counting metrics, descriptive economy can in fact be efficiently measured with help of the Information Theoretic notion of DESCRIPTION LENGTH.

Description length is a notion that can be evaluated on an implemented large-scale description using the notion of Kolmogorov complexity as shown by Sagot and Walther (2011). Our study compared the description lengths of four descriptions of (written) French making use of 1 to 139 inflection classes, various numbers of stem realization zones, and various degrees of stem suppletion. Among the four descriptions, one, extracted from the morphological encoding in the French electronic lexical resource (Sagot, 2010) is a fairly traditional one. It is similar to the description of French inflection along the lines of the standard conjugation dictionary Bescherelle (Arrivé, 1997) and makes use of 92 inflection classes. It is referred to as ORIG. A competing description, called BOBO, is an adaptation to written French of the account of French verbal inflection given by Bonami and Boyé (2003) using only one inflection class and a high degree of stem suppletion. Sagot and Walther developed a third description as an artificial counterpart to BOBO: it has 139 inflection classes and no stem allomorphy whatsover. This means in particular that a verb like ALLER go (see example (14)) has an empty stem while the entire word form is represented by its particular set of verbal endings. This third description is referred to as FLAT. The fourth description is one using realization zones and combining them into 20 inflection classes while also stipulating up to 12 suppletive stems for some verbal lexemes. This last description has been labeled NEW.

After implementing the four analyses of the verbal data from the large scale lexicon (7,000 lemmas realizing 360,000 word forms) using PARSLI's implementation, we measured the length of each one of the four descriptions using

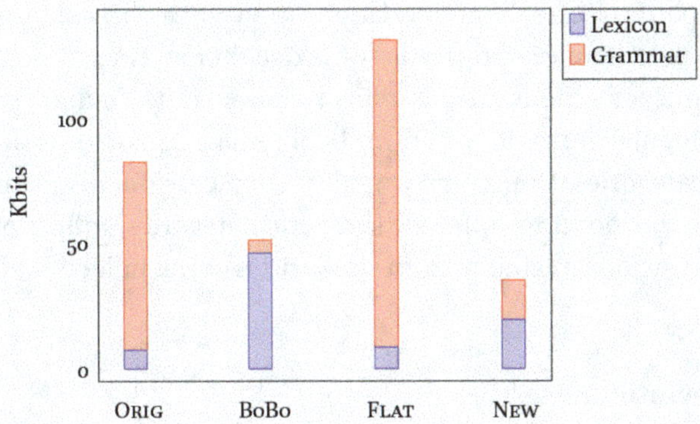

FIGURE 7.7 *Description length of four competing accounts of French verbal inflection according to Sagot and Walther (2011).*

the notion of Minimum Description Length (MDL: Rissanen, 1984) based on the Kolmogorov complexity. The overall description length of the four descriptions, as well as the distribution of the information between the grammar and the lexicon, is illustrated in Figure 7.7. It shows that while positing a single inflection class as in BOBO leads to a more economical account than using the standard description with 92 inflection classes, a better distribution, as illustrated by NEW with an intermediate number of inflection classes, leads to even better results.

Such an evaluation method thus offers a quantifiable means for assessing the number of inflection classes in a language when looked at from a realizational perspective. It is because inflection classes are not a primary notion but are derived from primary pattern characteristics and frequency patterns, that PARSLI offers a way to objectively assess the number of inflection classes in a system.[25]

Finally, the same method could also be applied to identify the number of inflectional categories one would like to posit in a particular description. A PARSLI inflectional category is a set of lexical items sharing a set of defaults, namely a set of realized morphosyntactic feature bundles and a particular structure of their

25 It is however important to stress that in order to be sure achieve significant results, this method requires large scale implementations, i.e., an implementation comprising a large enough number of the higher frequency lexemes so that adding a new random (less frequent) one will not change the balance between regular and irregular patterns, but rather enforce the difference.

inflectional pattern. Like inflection classes, they are a derived notion, a generalization over paradigmatic properties of lexical items. Thus, if half of the Slovak nouns, and not only the few exceptions mentioned by Zauner (1973), displayed a vocative singular form, it would probably make sense to posit two different *inflectional* categories. Doing this would of course by no means imply that they would also correspond to different syntactic categories, inflectional categories simply reflecting observable pattern similarities at the inflectional level.

Conclusion

In this chapter, we have introduced PARSLI, a model of inflection within WP approaches that focuses mainly on the lexicon and its relationship to individual paradigm shapes.

The primary difference between current WP models of inflection and the PARSLI model probably lies in the fact that it adopts a 'bottom up' approach, where most available models are 'top down': In PARSLI, the structure of the system is seen as a generalization of individual lexemic properties. Therefore the descriptive primitives of the morphological system are built into the definition of the lexical entries as atomic properties defining dimensions of variation, while generalizations emerge from the similarities between them. This model thus differs from existing models such as, for example, PFM, in which the basic building blocks are not the overall paradigm shape of particular lexemes, but general form-feature correspondences on the one hand and the inventory of said features on the other. In particular, most other existing models take notions like INFLECTIONAL CLASSES and MORPHOSYNTACTIC PROPERTY INVENTORIES to be basic notions. Within PARSLI, they do however not constitute descriptive primitives but evaluable secondary generalizations.

PARSLI, as it is presented here, is limited to the representation of inflectional morphology from a realizational perspective. Extensions of PARSLI are however currently being developed. Among those, one extends the model to derivational morphology and develops its integration with an established lexicalist syntactic framework. A second extension tries to adapt PARSLI modeling of lexical properties to an implicational view of morphology, as developed, among others, by Blevins (2006, 2016), Ackerman et al. (2009), and Ackerman and Malouf (2013), focusing more narrowly on the representation of predictive structure of morphological paradigms and analogical relationships between individual word forms or sets of forms and taking into account the dynamic adaptive system properties of language.

References

Ackerman, F., J.P. Blevins, and R. Malouf (2009). Parts and wholes: Implicative patterns in inflectional paradigms. In J.P. Blevins and J. Blevins (Eds.), *Analogy in Grammar: Form and Acquisition*, pp. 54–81. Oxford: OUP.

Ackerman, F. and R. Malouf (2013). Morphological organization: The Low Conditional Entropy Conjecture. *Language 89*, 429–464.

Ackerman, F., G.T. Stump, and G. Webelhuth (2011). Lexicalism, periphrasis and implicative morphology. In R. Borsley and K. Börjars (Eds.), *Non-transformational Theories of Grammar*, pp. 325–358. Oxford: Blackwell.

Anderson, S.R. (1982). Where's morphology? *Linguistic Inquiry 13*, 571–612.

Aronoff, M. (1994). *Morphology by Itself: Stems and Inflectional Classes*. Linguistic Inquiry Monographs. MIT Press.

Arrivé, M. (Ed.) (1997). *Bescherelle: La conjugaison pour tous*. Paris: Didier Hatier.

Baerman, M. (2006, sept). Deponency in serbo-croatian. Online database: Typological Database on Deponency, Surrey Morphology Group, CMC, University of Surrey.

Baerman, M. (2007a). Morphological reversals. *Journal of Linguistics 43*, 33–61.

Baerman, M. (2007b). Morphological typology of deponency. In M. Baerman, G.G. Corbett, D. Brown, and A. Hippisley (Eds.), *Deponency and Morphological Mismatches*, Volume 145, Oxford, pp. 1–19. The British Academy: OUP.

Baerman, M., G.G. Corbett, and D. Brown (Eds.) (2010). *Defective Paradigms: Missing Forms and What They Tell Us*. Oxford: OUP.

Blevins, J.P. (2006). Word-based morphology. *Journal of Linguistics 42*(3), 531–573.

Blevins, J.P. (2016). *Word and Paradigm Morphology*. Oxford: OUP.

Boas, H.C. and I.A. Sag (2012, May). Sign-based construction grammar: An informal synopsis. In H.C. Boas and I.A. Sag (Eds.), *Sign-Based Construction Grammar*. Stanford, CA: CSLI Publications. Online draft from May 2012.

Bonami, O. and G. Boyé (2002). Suppletion and dependency in inflectional morphology. In F.V. Eynde, L. Hellan, and D. Beerman (Eds.), *The Proceedings of the HPSG '01 Conference*, Stanford, CA. CSLI Publications.

Bonami, O. and G. Boyé (2003). Supplétion et classes flexionnelles dans la conjugaison du français. *Langages 152*, 102–126.

Boyé, G. (2006). Suppletion. In K. Brown (Ed.), *Encyclopedia of Language and Linguistics (2nd ed.)*, Volume 12, pp. 297–299. Oxford: Elsevier.

Bresnan, J. (1982). *The Mental Representation of Grammatical Relations*. MIT Press.

Brown, D. and M. Chumakina (2012). What there is and what there might be. In D. Brown, M. Chumakina, and G. Corbett (Eds.), *Canonical Morphology and Syntax*, pp. 1–19. Oxford: OUP.

Brown, D., G.G. Corbett, N. Fraser, A. Hippisley, and A. Timberlake (1996). Russian noun stress and network morphology. *Linguistics 34*, 53–107.

Brown, D. and A. Hippisley (2012). *Network Morphology: A Defaults-based Theory of Word Structure*. Cambridge: CUP.

Chomsky, N. (1965). *Aspects of the Theory of Syntax*. MIT Press.

Corbett, G.G. (2007). Canonical typology, suppletion and possible words. *Language 83*, 8–42.

Corbett, G.G. and N. Fraser (1993). Network Morphology: a DATR account of Russian nominal inflection. *Journal of Linguistics 29*, 113–142.

Crysmann, B. and O. Bonami (2015). Variable morphotactics in information-based morphology. *Journal of Linguistics 52*, 311–374.

Dalrymple, M. (2001). *Lexical Functional Grammar*, Volume 34 of *Syntax and Semantics*. Academic Press.

Evans, R.P. and G. Gazdar (1989). Inference in DATR. In *Proceedings of the Fourth Conference of the European Chapter of the Association for Computational Linguistics*, pp. 66–71.

Evans, R.P. and G. Gazdar (1996). DATR: A language for lexical knowledge representation. *Computational Linguistics 22*, 167–216.

Fradin, B. and F. Kerleroux (2003). Troubles with lexemes. In G. Booij and A.R. Janet de Cesaris, Sergio Scalise (Eds.), *Selected papers from the Third Mediterranean Morphology Meeting*, Topics in Morphology, Barcelona, pp. 177–196. IULA-Universitat Pompeu Fabra.

Hippisley, A. (1997). *Declarative Derivation*. Ph. D. thesis, University of Surrey.

Hockett, C.F. (1954). Two models of linguistic description. *Word 10*, 210–234.

Kiparsky, P. (1982). Lexical morphology and phonology. In I.-S. Yang (Ed.), *Linguistics in the Morning Calm: Selected Papers from SICOL*, pp. 3–91. Seaoul: Hanshin.

Koenig, J.-P. (1999). *Lexical Relations*. Stanford Monographs in Linguistics. Stanford, CA: CSLI Publications.

Lyons, J. (1963). *Structural Semantics: An Analysis of Part of the Vocabulary of Plato*, Volume 20 of *Publications of the Philological Society*. Oxford: Blackwell.

Matthews, P.H. (1965). The inflectional component of a word-and-paradigm grammar. *Journal of Linguistics 1*(2), 139–171.

Matthews, P.H. (1972). *Inflectional Morphology: a Theoretical Study Based on Aspect of Latin Verb Conjugation*. Cambridge: CUP.

Matthews, P.H. (1974). *Morphology*. Cambridge: CUP.

Pirrelli, V. and M. Battista (2000). The paradigmatic dimension of stem allomorphy in italian verb inflection. *Italian Journal of Linguistics 12*, 307–380.

Pollard, C. and I. Sag (1994). *Head-driven Phrase Structure Grammar*. Stanford, CA: CSLI Publications.

Rissanen, J. (1984). Universal coding, information, prediction, and estimation. *IEEE Transactions on Information Theory 30*(4), 629–636.

Robins, R.H. (1959). In defense of WP. *Transactions of the Philological Society 1959 58*(1), 116–144.

Sagot, B. (2010). The *Lefff*, a freely available, accurate and large-coverage lexicon for French. In *Proceedings of the 7th Language Resource and Evaluation Conference*, Valletta, pp. 2744–2751.

Sagot, B. (2014). DeLex, a freely-avaible, large-scale and linguistically grounded morphological lexicon for German. In *Proceedings of the 9th Language Resources and Evaluation Conference*, Reykjavik, pp. 2788–2784.

Sagot, B. and G. Walther (2011). Non-canonical inflection: data, formalisation and complexity measures. In C. Mahlow and M. Piotrowski (Eds.), *Systems and Frameworks in Computational Morphology*, Volume 100, Zurich, pp. 23–45. Springer.

Sagot, B. and G. Walther (2013). Implementing a formal model of inflectional morphology. In C. Mahlow and M. Piotrowski (Eds.), *Systems and Frameworks in Computational Morphology*, Volume 100, Berlin, pp. 115–134. Springer.

Spencer, A. (2005). Generalized Paradigm Function Morphology—A synopsis. In A. Galani and B. Sinar (Eds.), *York Papers in Linguistics, Papers from the York-Essex Morphology Meeting, 2003*, Volume 2, pp. 93–106. University of York.

Spencer, A. (2013). *Lexical Relatedness: A Paradigm-based Model*. Oxford: OUP.

Stump, G.T. (2001). *Inflectional Morphology. A Theory of Paradigm Structure*. Cambridge: CUP.

Stump, G.T. (2006). Heteroclisis and paradigm linkage. *Language 82*, 279–322.

Stump, G.T. (2014, May). Polyfunctionality, positional exponence and affix conflation. Keynote presentation at the 16th International Morphology Meeting, Budapest, Hungary.

Thornton, A.M. (2011). Overabundance (multiple forms realizing the same cell): A non-canonical phenomenon in italian verb morphology. In M.G. Martin Maiden, John Charles Smith and M.-O. Hinzelin (Eds.), *Morphological Autonomy: Perspectives From Romance Inflectional Morphology*, pp. 358–381. Oxford: OUP.

Walther, G. (2011). Measuring morphological canonicity. In G. Perko (Ed.), *Linguistica*, Volume 51, pp. 157–180. Ljubljańa: Faculty of Arts, University of Ljubljana. Internal and External Boundaries of Morphology.

Walther, G. (2013). *Sur la canonicité en morphologie—Perspective empirique, formelle et computationnelle*. Ph. D. thesis, Université Paris-Diderot.

Zauner, A. (1973). *Praktická príručka slovenského pravopisu*. Martin: Vydavateľstvo Osveta.

CHAPTER 8

The Possible Role of Entropy in Processing Argument Dependencies in Hungarian

Csaba Pléh, István Fekete, and Dániel Varga

It has been commonplace in the research of sentence processing for at least 40 years now, since the beginning of the use of Thematic Roles and different versions of Frame Theories in processing studies, to find facilitative effects between verbs and their arguments (Tanenhaus et al, 1995, Kintsch, 1998). The basic idea was proposed in a rather simple manner by early frame and schema based theories of understanding (Schank, 1972, Kintsch, 1994). Understanding a sentence involves two basic stages:

1. Activating the representation of predicates from a long-term store, together with their argument frames, expected and likely arguments.
2. Filling the arguments with actual phrases from the string.

This kind of approach has been supported many times both theoretically and experimentally by postulating and proving priming relations from processing the predicate towards processing the arguments. Thus reading the verb CUT would facilitate the reading of instrumental arguments (Kinsch, 1998, Kintsch & Mangalath, 2011). There were many discussons over the four decades about the relative impact of general expectations (CUT INSTR) and specific lexical expectations (CUT knife).

These early cognitive theories—which are still continued by the group of Kintsch—were not interested in what grammatical markers are used by the system to assign argument roles to certain noun phrases. This was changed by a more detailed linguistic shift towards the argument relations, with the idea that Thematic Roles are keys to the syntax-semantic interface in understanding (Carlson and Tanenhaus, 1988, Tanenhaus, Carlson, and Trueswell, 1989). Several studies have shown that arguments are processed faster than adjuncts (Kennison, 2002), as adjuncts are optional as opposed to arguments, and that assignment ambiguities slow down processing, such as the structural ambiguity in (1), which is more difficult to process than (2) because either John or the policeman could have the *binoculars*, whereas in (2) it can only be the policeman that carries the *gun*.

 | DOI 10.1163/9789004342934_010

(1) *John saw the policeman with the binoculars.*
(2) *John saw the policeman with the gun.*

The step of Thematic Role assignment has started to play a central role in syntax based parsing theories (Frazier and Fodor, 1978, Ferreira and Clifton, 1986), with much discussion on the automaticity and modularity issues. These considerations have introduced the issue of the possible role of morphology in these expectation-based processes.

Bornkessel and Schlesewsky (2006, Bornkessel-Schlesewsky et al., 2011) have run many behavioral, evoked potential, and imaging studies on sentence understanding in languages using different types of cues to argument roles (order, animacy, case marking). Independently of the cues used, Broca's area always played a crucial role in assigning Thematic Roles. On the basis of these processing data, Bornkessel & Schlesewsky (2006) developed a fully-fledged cross-linguistic theory of the temporal activation of the verbal argument frames and the insertion of noun phrases into the slots as a second step based on neuronal processing evidence.

In our work we concentrate on oblique arguments of Hungarian verbs exploiting the fact that argument relations are coded by case markers in Hungarian (Kiefer, 1987, 2003). Argument processing and Thematic Role assignment in such a language is closely tied with morphological processing. In particular, most arguments with abstract relations are coded by concrete/spatial case markers that originally denoted locational relationships in the physical world. For example, the abstract relation in the Hungarian sentence *János fél a kutyától* ('John is afraid of the dog') is expressed by the ablative suffix *-tól/-től* ('from'). The suffix has two realizations (*-tól/-től*), determined by the phonological environment of the noun stem to which it is attached. This alternation of physical/concrete and abstract argument use results several times in ambiguities to decide whether a given NP is an abstract argument or a concrete adjunct. The alternation between spatial and abstract argument interpretation often results in ambiguities.

There has been much discussion on how to differentiate arguments from adjuncts in Hungarian (Komlósy, 1994, Alberti et al, 2014). Table 8.1 illustrates some of these intricacies.

There are, of course, ambiguous argument frames in English too. The sentence *John decided on the boat* can be interpreted either as John chose the boat or that he made his decision while on the boat (Hornstein and Weinberg, 1981). However, in English these cases result from structural ambiguity based on attachment height of the prepositional phrase (PP; in

TABLE 8.1 *Varieties of the morphology of some argument frames in Hungarian*

Type	Example	Gloss
Unambigouous	Emlékszik a fi**úra** Haragszik a tanít**óra.** Készül a versen**yre.** Találkozott a lá**nnyal.** Csókolózott a rendő**rrel.**	remembers the boy-ON angers the teacher-ON Prepares the race-ON Met the girls-INST Kissed the cop-INST
ambiguous in marking	Gondol a lány**ra.** Gondol valamit.	thinks girl-ON thinks something-ACC
ambiguous argument / adjunct	Gondolkodik a lány**on.** Gondolkodik a haj**ón.**	thinks girl-ON thinks boat-ON

the cited example the PP is 'on the boat'). In Hungarian, however, they are related to the case suffixes of nouns. Earlier studies on the processing of argument structures in Hungarian have shown that these assumed interactions between morphology and sentence processing do indeed hold. Gervain and Pléh (2004) showed that prenominal, sentence initial verbs facilitate the processing of constructions like 'Anna thought of the boat,' and postverbal nouns that are ambiguous between a locational and an abstract-argument reading are read slower than ordinary arguments. Compare 'Anna RUMINATED on the boat' versus 'Anna RUMINATED on the problem,' where in the latter case the locative meaning is excluded. Gervain and Pléh interpreted their data as supporting priming, facilitative effects from the verb to morphological endings coding for arguments. Figure 8.1 shows one of the examples for the concrete/physical meaning being slower, as it activates the abstract meaning too.

Thus, the data imply a processing model where a verb based expectation arrow would metaphorically 'point' towards the argument. That is why the argument noun would be read faster compared to the adjunct/argument Ambiguity between locative and argument reading slows down reading the NP.

Instrumentals are interesting structures regarding both the possible arguments frames of the main verbs, warrying between comitative, and 'real instrumental' readings. They also alternate between argument and adjunct readings. (3) to (5) show some of the interesting structures we worked with.

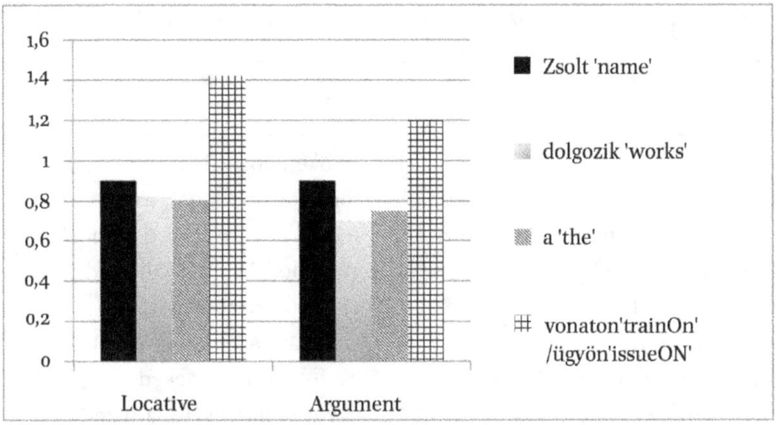

FIGURE 8.1 *Spatial case endings are ambiguous, thus they are read slower than arguments with the same case ending* 'John works on the train/ on the case.'

(3) *A katona a kert-ben* ***sétált*** *a lánny-*[1]*al.* **adjunct**
'The soldier the garden-IN walked the girl-INST'
(4) A *katona a kertben* ***játszott*** *a lánnyal.* **ambiguous**
'The soldier the garden-IN played the girl-INST'
(5) *A katona a kertben* ***találkozott*** *a lánnyal.* **argument**
'The soldier the garden-IN met the girl-INST'

Fekete and Pléh (2011) compared unidirectional (6), and bidirectional (7) comitative constructions, and the singular or plural anaphoric continuations in a word-by-word self paced reading paradigm.

(6) *A fiú gúnyolódott a lánnyal, aztán leült/leültek.*
'The boy mocked the girl-INST, and then sat down/and then they sat down.'
(7) *A fiú játszott a lánnyal, aztán leült/leültek.*
'The boy played the girl-INST, and then sat down/and then they sat down.'

By unidirectional comitative construction we mean an event with an agent and a patient, while a bidirectional construction involves an agent and a co-agent. Crucially, unidirectional constructions are semantically irreversible, while bidirectional constructions are reversible. Hungarian, as being a

1 Notice that there is fusional allomorphy here. Instrumental case *-val/-vel* is fused with stem ending consonats. Thus *lány-val* is realized as *lánnyal.* See Kiefer (2003).

TABLE 8.2 *The critical verbs used in the experiments on* COM *and* INST *processing*

Comitative	Instrumental
borozott = drank wine with	csipkelődött = bantered with
találkozott = met with	foglalkozott = dealt with
tegeződött = were on familiar terms with	incselkedett = joshed with
csókolózott = kissed with	gúnyolódott = japed/jested with
sétálgatott = walked ith	orított = shouted with
bulizott = partied with	szimpatizált = liked somebody
mulatozott = racketed with	csúfolódott = mocked
párbajozott = duelled with	együttérzett = sympathized with
verekedett = fought with	kikezdett = made a pass at sb.
énekelt = sang with	törődött = cared for someone

PRO-DROP language, uses zero anaphors in the continuation sentences in these constructions (Pléh and Radics, 1978). Table 8.2 shows the verbs used in these studies.

Not surprisingly, after unidirectionals (6), the singular, after bidirectionals (7), the plural continuations were read faster, showing a 'deep anaphora effect' in the sense introduced by Hankamer & Sag (1976). In other words, it is easier to refer to the subject of a unidirectional construction than referring to both the agent and the patient with the plural. However, in the case of a bidirectional construction, the plural reference is more accessible because the two agents are equal in terms of their involvement in the event.

Interestingly, reading times and answering times were both faster after the unidirectional instrumental frames. This is presumably due to the argument versus adjunct status of the verb-NP relations. The reading times are summarized on Figure 8.2. NVN (noun-verb-noun) indicates the word order used in this condition. Crucially, during sentence processing on the region of the verb the second argument is not spelled out, so in the bidirectional (comitative) condition the parser encounters a bifurcation, which causes a slowdown in processing (complexity spill-over): the sentence could as well finish as a grammatical sentence without an argument (e.g., the boy sang, partied, walked, etc.), or the verb could select for a second argument, a co-agent, while in the unidirectional (instrumental) condition the verb obligatorily selects for an argument, hence the faster processing times for the second arguments in the instrumental condition.

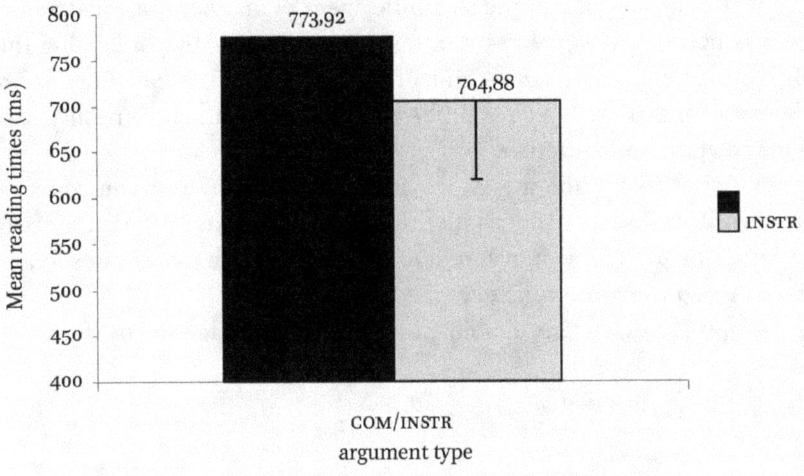

FIGURE 8.2 *Mean reading times of the arguments after the critical verbs (NVN) in the COM-INSTR experiment.*

Reanalysis with the Possible Role of Entropy

Although in the studies mentioned above the frequencies were controlled, and the role of collocations was controlled for by reversing the noun phrases in the different conditions, the issue of the statistical conditional predictability in the argument relations was not raised at all. However, it should be noted that we strived to step around this confound by conducting a separate experiment where the verbs were presented in sentence-final position, which rules out the possibility that verbs project their argument structure, that is, expectations are made about the continuation with a second argument (Fekete and Pléh, 2011). Importantly, Hungarian has a relatively free word order. This way, we can rule out the argument-adjunct effects, which arose in the previous condition where sentences were presented in NVN word order.

In the present paper, a post hoc reanalysis of the reaction time data of Fekete and Pléh (2011) is conducted by using entropy of relations as a possible predictor. Importantly, we are examining reaction time data of our experiment in which verbs presented were between two case marked NPs, thus between possible arguments. Informally, in this situation entropy is the uncertainty of the case ending following a given verb. So, for example, the Hungarian verb *harcol* ('fight') can select for three different oblique case endings (*-ért* 'for,' *-val/-vel* 'with,' *ellen* 'against') with unequal probability. Moreover, the second case ending 'with' is in itself ambiguous. The given actor can either refer to a co-agent who is on the side of the semantic agent or an enemy against whom the agent is fighting. This creates a rather uncertain situation following this verb. By contrast, the verb *gondol* 'thinks' licenses only one case ending, *-ra/-re* (literally 'on'). Thus there is a clear difference in terms of the uncertainty with regard to the argument structure.

To operationalize argument related conditional entropy, one can conceive of it as a measure of uncertainty about the nominal contexts in the presence of a given verb. Due to the free movement of NPs in Hungarian, both subsequent and preceding contexts are considered.

The entropy of a discrete probability distribution W is defined by (8).

(8) $$H(W) = -\sum_{w \in W} p(w)\log_2 p(w)$$

Consider the following simple random process: for a given verb word form, we pick a noun phrase in the corpus within a predetermined window form this verb, and take the case of this noun phrase. We define the conditional entropy of the case frames of a verb as the entropy of the output of this random process. Intuitively, it measures how diverse the possible set of cases accompanying a verb is, giving less weight to rare cases.

We computed entropy measures from two corpora. From the MOKK (2006, Kornai et al, 2006) and the MAZSOLA corpus (Sas, 2008) entropy computations were made for the relations between a given verb and the noun endings in a 'plus/minus two content words' frame. These entropy estimates will be used as predictors in the self-paced word-by-word reading time experiments of Fekete and Pléh for the reading of nouns in sentence contexts. In this we wanted to learn whether the entropy relations between verbs and case endings do have an explanatory power in processing and verb-noun attachments.

Entropy in the Two Corpora

The *Szószablya* corpus does not identify dependents, so we work with nouns following the verb within a fixed (4-token) window. The *Mazsola* corpus does identify dependents, but its size is somewhat smaller. (226M tokens versus 712M tokens.)

In the case of the *Mazsola* corpus, we only required the verb stem to correspond, thus, citation forms were used, while in the case of the *Szószablya* corpus, we required the exact word form of the verb as used in the experiments.

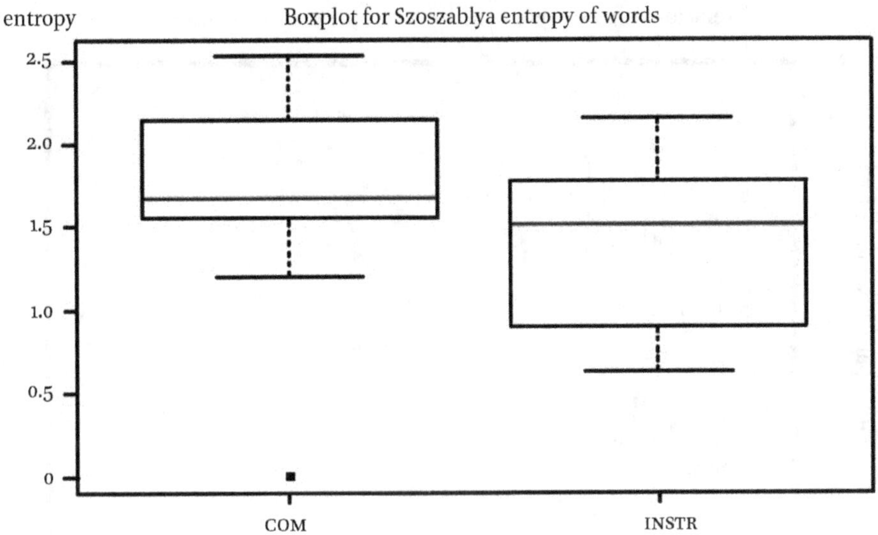

FIGURE 8.3 *Entropy of COMIT is higher.*

The correlation between the two entropy measures was extremely high, 0.937, showing that the entropy measure from the two corpora are robustly similar.

The comitative, bidirectional verbs on the whole had a less predictable relation to the nominal case endings in their environment, presumably due to the adjunct status of the oblique NP. As Figure 8.3 shows, their entropy was much higher.

There was a not simple but interesting relationship between the reaction time differences in the verb pairs and entropy, as Figure 8.4. shows
It is important to note that although the reading times above denote the reading times of word 5, this reading time reflects sentence integration processes, given that this position is the sentence-final position.

Conclusions

- The entropy notion extended to Verb-Argument relations is rather robust
- It is related to argument frame differences such as INSTR COMIT differences
- The entropy of the argument frame has some effects on processing speed

The ambiguity of argument frames that was originally used in these studies can be conceived of as a situation of maximum entropy when all possible argument

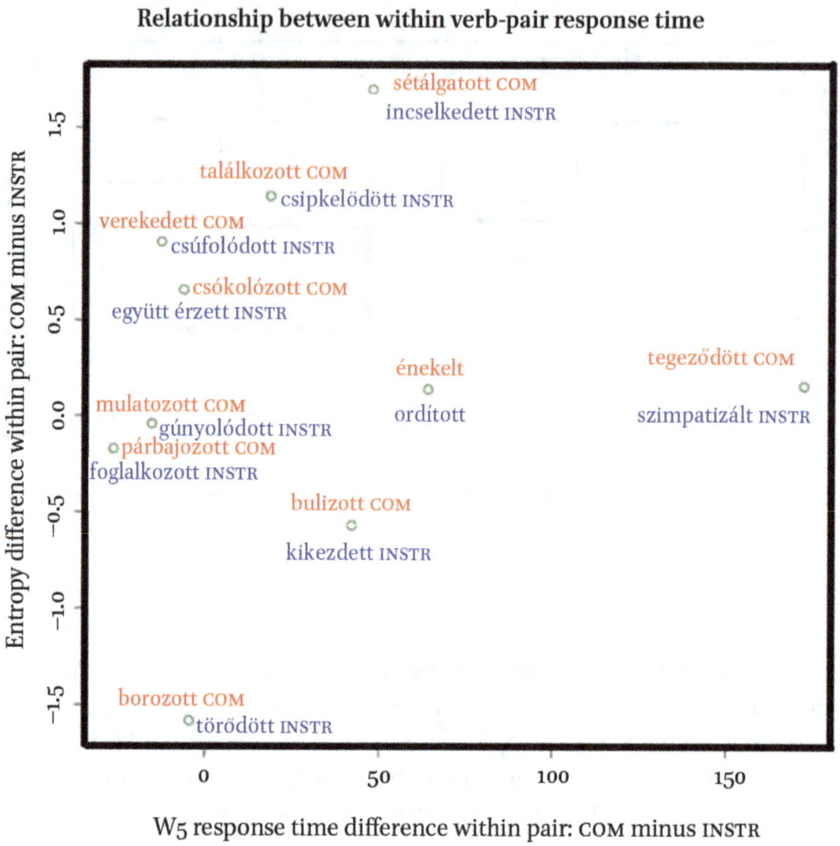

FIGURE 8.4 *Entropy differences correspond to RT differences.*

frames are equally likely. Its relation to grammatical effects and semantic priming still has to be clarified. It could be the case that the deep anaphora effect revealed in Fekete and Pléh (2011) is a spillover effect consistent with the difference in entropy between the two case conditions, rather than resulting from a difference in mental representations of the scenarios described by the sentences. In other words, differences in verb-conditioned entropy could in principle explain the resulting pattern more parsimoniously than the putative representational-semantic difference between comitative and instrumental constructions.

Language understanding, which involves both bottom-up and top-down (contextual) processes, is not only inductive but also expectation-driven. We are faster at processing language material which has been anticipated, predicted, or highly probable based on the context or the linguistic environment in a sentence (e.g., Balota, Yap, & Cortese, 2006). One of these expectation-driven processes is the projection of the argument structure of verbs (the activation of candidate case frames in Hungarian), which is an automatic, very rapid and unconscious process contingent on the entropy of the verb in question. The different psycholinguistic factors such as word frequency, word length, affix type frequency, etc. jointly determine processing time. One of these influencing factors that can predict reading times is the entropy of verb related argument structure.

Our results are in line with findings in the field of inflectional entropy. Moscoso del Prado Martín, Kostic, and Baayen (2004), for example, demonstrated that visual lexical decision latencies of Dutch nouns were positively correlated with inflectional entropy. Similarly to the entropy of argument structure, inflectional entropy indicates the uncertainty as to the probability and the number of inflectional variants the noun has. Specifically, it is higher for nouns that have more inflectional variants as opposed to nouns that have fewer. Crucially, nouns with inflectional variants of equal or similar probabilities also result in an entropy increase. Pylkkänen et al. (2004) showed that such an effect can be shown at the neuronal level too.

Entropy might provide better explanations for some psycholinguistic phenomena than traditional factors taken into account. Technically this suggests that entropy as a sensitive psycholinguistic variable should be taken into consideration in experimental psycholinguistics by gathering entropy measures from morphosyntactically-annotated corpora or psycholinguistic norms.

References

Alberti, G., Gervain, J., Schnell, Zs., Szabó, V. & Tóth, B. (2014). A vonzatsorrend és az esetmorfológia külsö meghatározottsága" [The external determination of argument order and case morphology]. In: Kádár, Edit and Szilágyi N., Sándor (eds.): *Motiváció és modern nyelvleírás* [Motivation and contemporary language theory]. Cluj Napoca.

Balota, D.A., Yap, M.J., & Cortese, M.J. (2006). Visual word recognition: The journey from features to meaning (A travel update). In M. Traxler & M.A. Gernsbacher (Eds.) Handbook of psycholinguistics (2nd edition) (pp. 285–375). Amsterdam: Academic Press.

Bornkessel, I. & Schlesewsky, M. (2006). The extended argument dependency model: A neurocognitive approach to sentence comprehension across languages. *Psychological Review, 113*(4), 787–821.

Bornkessel-Schlesewsky, I., Kretzschmar, F., Tune, S., Wang, L., Genç, S., Philipp, M., Roehm, D., Schlesewsky, M. (2011). Think globally: Cross-linguistic variation in electrophysiological activity during sentence comprehension. *Brain and Language, 117*, 133–152.

Carlson, G. & Tanenhaus, M. (1988). Thematic roles and language comprehension. *Syntax and Semantics. Vol. 21*, 263–288.

Fekete, I. & Pléh, Cs. (2011). Bidirectional and unidirectional comitative constructions in Hungarian: A psycholinguistic investigation at the interface of argument structure and semantics. *Acta Linguistica Hungarica, Vol. 58 (1–2)*, pp. 3–23.

Ferreira, F. & Clifton, C. (1986). The independence of syntactic parsing. *Journal of Memory and Language, 25*, 348–368.

Frazier, L & Fodor, Janet (1978). The sausage machine: A new two-stage *parsing* model. *Cognition, 6*, 291–325.

Gervain, J. & Pléh, C. (2004). [Anna is thinking on the ship, or argument expectancies in Hungarian sentence understanding]. In: Judit Gervain, J. & Pléh, C. (eds): *A láthatatlan megismerés*, 'Invisible cognition', 112–25. Gondolat, Budapest. In Hungarian.

Hankamer, J. & Sag. I. (1976). Deep and surface anaphora. *Linguistic Inquiry 7*: 391–428.

Hornstein, N. & Weinberg, A. (1981). Case Theory and Preposition Stranding. *Linguistic Inquiry, 12*, 55–91.

Kennison, S.N. (2002). Comprehending Noun Phrase Arguments and Adjuncts. *Journal of Psycholinguistic Research, 31, 65–81.*

Kiefer, F. (1987). The *cases of Hungarian nouns.* Acta Linguistica Academiae Scientiarum, 37, 93–101.

Kiefer, F. (2003). Alaktan (Morphology). In: É. Kiss Katalin, Kiefer Ferenc, Siptár Péter, *Új magyar nyelvtan.* 'New Hungarian grammar'. Osiris, Budapest, 2003, 189–284.

Kintsch, N. (1974). *The representation of meaning in memory.* Hillsdale: Erlbaum.

Kintsch, W. (1998). *Comprehension: A paradigm for cognition.* Cambridge, UK: Cambridge University Press.

Kintsch, W. & Mangalath, P. (2011). The construction of meaning. *Topics in Cognitive Science, 3*, 346–370.

Komlósy, A. (1994). Complements and Adjuncts. In Kiefer, Ferenc & É. Kiss, Katalin eds. *The Syntactic Structure of Hungarian* (= Syntax and Semantics 27). San Diego: Academic Press. 91–178.

Kornai, A. et al. (2006). Web-based frequency dictionaries for medium density languages, In *Proceedings of WaC'06, Szószablya corpus*, http://szotar.mokk.bme.hu/szoszablya/searchq.php.

MOKK (2006). A WEB based Hungarian frequency dictionary: http://szotar.mokk.bme.hu/szoszablya/searchq.php.

Moscoso del Prado Martin, F., Bertram, R., Schreuder, R., and Baayen, R.H. (2004). Morphological size in a morphologically rich language: The case of Finnish compared to Dutch and Hebrew, *Journal of Experimental Psychology: Learning, Memory, and Cognition, 30*, 1271–1278.

Pléh, Cs., Radics, K. (1976). 'Hiányos mondat,' pronominalizáció és a szöveg" ["Elliptical sentence," pronominalizaton and text]. *Általános Nyelvészeti Tanulmányok* [*Studies in General Linguistics*], *9*, 261–277.

Pléh, Cs. , and Radics, K. (1978). "Truncated sentence, pronominalization and the text". *Acta Linguistica Academiae Scientiarum Hungan'cae, 33*, 91–113.

Sass, B. (2008): The Verb Argument Browser. In: Sojka, P. et al. (eds.): Proceedings of TSD 2008, Brno, Czech Republic, LNCS 5246, 187–192, http://corpus.nytud.hu/vab.

Schank, R. (1972). Conceptual dependency: A theory of natural language understanding. *Cognitive Psychology*, *3*, 552–631.

Tanenhaus, M.K., Carlson, G., and Trueswell, J.C. (1989). The role of thematic structures in interpretation and parsing. *Language and Cognitive Processes, 4*, 211–234.

Tanenhaus, M.K., Spivey-Knowlton, M.J., Eberhard, K., & Sedivy, J.C. (1995). Integration of visual and linguistic information in spoken language comprehension. *Science, 268*, 632–634.

CHAPTER 9

A Functional Theory of Gender Paradigms

Melody Dye, Petar Milin, Richard Futrell, and Michael Ramscar

In German ... every noun has a gender, and there is no sense or system in the distribution; so the gender of each must be learned separately and by heart. There is no other way. To do this one has to have a memory like a memorandum-book. In German, a young lady has no sex, while a turnip has. Think what overwrought reverence that shows for the turnip, and what callous disrespect for the girl ... :

Gretchen: Wilhelm, where is the turnip?
Wilhelm: She has gone to the kitchen.
Gretchen: Where is the accomplished and beautiful English maiden?
Wilhelm: It has gone to the opera.

... a tree is male, its buds are female, its leaves are neuter; horses are sexless, dogs are male, cats are female—tomcats included, of course; a person's mouth, neck, bosom, elbows, fingers, nails, feet, and body are of the male sex, and his head is male or neuter according to the word selected to signify it, and **not** according to the sex of the individual who wears it—for in Germany all the women wear either male heads or sexless ones; a person's nose, lips, shoulders, breast, hands, and toes are of the female sex; and his hair, ears, eyes, chin, legs, knees, heart, and conscience haven't any sex at all. The inventor of the language probably got what he knew about a conscience from hearsay.

MARK TWAIN, (1880) *"The Awful German Language"*

•••

The confusions that occupy us arise when language is like an engine idling, not when it is doing work.

WITTGENSTEIN (1953), *Philosophische Untersuchungen*

 | DOI 10.1163/9789004342934_011

1 Introduction

In his humorous account of the "awful" German language, Mark Twain draws attention to a puzzle posed by many of the world's languages: grammatical gender. As often as not, the languages of the world assign objects into seemingly arbitrary (and often seemingly sexist) noun classes that lack any transparent purpose (Corbett, 1991). Historically, this led some scholars to conclude that grammatical gender is senseless: William of Ockham considered gender to be a meaningless, unnecessary aspect of language, an obvious candidate for his famous razor; Baudouin de Courtenay described gender as a deformity, an unfortunate historical accident that was responsible for a range of human afflictions, including nightmares, pathological behavior, erotic and religious delusions, and sadism (Kilarski, 2007). Few other linguists have held noun class to be responsible for all of the world's ills; but few have warmed to its virtues either. The consensus is neatly summarized by Leonard Bloomfield (1933): "[t]here seems to be no practical criterion by which the gender of a noun in German, French, or Latin [can] be determined."

Not only have gender systems been branded as meaningless, but they are fiendishly difficult for non-native speakers to learn, a state of affairs that prompted the developmental psychologist Michael Maratsos (1979) to conclude:

> The presence of such systems [German gender] in a human cognitive system constitutes by itself excellent testimony to the occasional nonsensibleness of the species. Not only was this system devised by humans, but generation after generation of children peaceably relearns it.

While many linguists have reconciled themselves to the idea that gender has evolved its negative consequences for no reason, Charles Darwin was less sanguine about such matters: "The sight of a feather in a peacock's tail, whenever I gaze at it, makes me *sick*," he famously wrote.[1] In the 1800's, Darwin's pursuit of evolutionary explanations for such apparent anomalies revolutionized our understanding of biology. Indeed, his ruminations on the peacock's tail helped develop the theory of sexual selection: Darwin hypothesized that while the extravagance of the male peacock's train might prove hazardous to its health, females would often opt for mates with more ornate plumage, leading to reproductive success for showier males. Hence even the seemingly 'absurd' and risky feather display of a male peacock might still have an adaptive purpose. In this

1 Letter 2743—Darwin, C.R. to Gray, Asa, 3 Apr (1860), Darwin Correspondence Project.

chapter, we adopt Darwin's stance in analyzing the place of grammatical gender in German, seeking to elucidate a functional role for gender marking in facilitating communicative efficiency.

1.1 *Some Proposed Functions of Noun Class*

Grammatical gender is an obligatory morphological system found in many languages that groups nouns into a small number of mutually exclusive classes, and marks neighboring words (such as articles and adjectives) for agreement. Many languages, such as French and Spanish, divide nouns into two distinct classes: masculine and feminine. Others, like German and Russian, add a third neuter category, yet even more are possible; Swahili has six (Corbett, 1991). Speaking broadly, the gender of nouns tends to be semantically arbitrary, such that there is little obvious correspondence between the conceptual properties of a referent and its noun class, and there is substantial cross-linguistic variation in which nouns get classed together (Vigliocco et al., 2005).

While not all researchers consider noun class to be purely ornamental, many of the functions that have been proposed for other languages have only limited applicability in German, the focus of the present study. One hypothesis is that gender marking assists comprehension processes by linking temporally separated elements in discourse, establishing local and global coherence. For instance, in some languages—but not German—agreeing gender markers can facilitate freedom in word order by marking which words describe the same thing. As can be seen in (1), Latin 'attributive' adjectives need not appear in a fixed position relative to nouns because suffixes are declined for gender, case, and number, and this helps make clear which adjectives and nouns belong together:

(1) ultim-a Cumae-i ven-it iam carmin-is aet-as
last-NOM.FEM Cumai-GEN.NEU came now song-GEN.NEU age-NOM.FE
"The last age of the Cumaean song has now arrived."[2]

Yet German does not have this functionality. Only attributive adjectives are marked for agreement, and those adjectives cannot appear anywhere other than immediately before a noun:

(2) die *große* Frau sah das Kind

(3) *die Frau sah das Kind *große*

2 From Virgil's Aeneid, cited in Matthews 1981 and Evans 2010.

Perhaps the most concrete suggestion that has been put forward for German gender's function, is that agreement between gender markers and anaphoric pronouns facilitates reference tracking (Zubin & Köpcke, 1986; Koval, 1979; Heath, 1975, *i.a.*). Consider the following:

(4) der Krug fiel in die Schale, aber er zerbrach nicht
the.MAS jug fell into the bowl.FEM but it.MAS broke not
'The jug fell into the bowl, but it (the jug) didn't break.'

In this instance, the referent of the pronoun 'it' is unambiguous, because 'it' must have a MASCULINE referent (which in this case must be the jug, not the bowl). However, even this proposal suffers shortcomings. For one, the existence of semantic regularities in noun class works *against* reference tracking, by increasing the probability that confusable nouns will be referenced with the same gendered pronoun (Lakoff, 1986). For another, German grammar frequently does not permit its speakers to rely on gender for this kind of discrimination (Claudi, 1985).

As these examples illustrate, gender may play different roles in different languages. Indeed, there is a growing body of evidence attesting to cross-linguistic differences in morphosyntactic processing, showing substantial variation in how listeners make use of gendered determiners in discourse (see e.g., Miozzo & Caramazza, 1999; Schriefers & Teruel, 2000). Accordingly, it would be a mistake to treat all systems called "noun class" as the same thing and to ignore the details of how, when, and where language speakers mark gender (see also MacWhinney, Bates, & Kliegel, 1984).

In determining the function of noun class in a given language, it is critical to examine the part that gender marking plays both in communication between current speakers (information processing) and in transmission between generations (learning). In what follows, we conduct precisely such an examination from the vantage point of information theory. While information theory is typically considered in the context of modern computing and engineering, it provides a useful lens through which to consider human language. In particular, its mathematical toolkit offers a precise means of quantifying how information is distributed across a language. By measuring systematic variations in that distribution in German, we are able to investigate how gendered determiners aid efficiency in linguistic processing. The findings we present here provide compelling support for the idea that grammatical gender is no mere ornament. On the contrary, gender appears to be an invaluable resource for regulating the flow of information between speakers.

2 An Information Theoretic Approach

To understand the importance of information theory to this problem, it is useful to contrast the lens it offers against that of the standard linguistic model. Since antiquity, language has mainly been conceived of in terms of a single, dominant metaphor: that of the direct material exchange of messages. According to the 'conduit' of metaphor of communication, a speaker packs the content of a message into words, which a listener unpacks at the other end. Utterances are supposed to somehow 'contain' their meanings, much as a stamped envelope contains a letter (Reddy, 1979). This metaphor for understanding language is pervasive in folk psychology, and is reflected in a broad array of psychological and linguistic theories (Sperber & Wilson, 1996).

Yet the conduit metaphor is neither inevitable nor irresistible. Conveyance systems, such as the mail or the carrier pigeon, are not the only means by which human societies have communicated at a distance, and an indirect alternative, in which messages are telegraphed across space and time, rather than physically conveyed by transport, has long been available. In telegraphy, no material copy of the message is ever sent. Instead, the message is translated into a physical signal that can travel the distance required. Successful communication relies on both the sender and receiver sharing the same code, such that the receiver can discriminate the original message from the received signal (Holzmann & Pehrson 1994). Modern digital communications are conceived of within precisely this framework, which Shannon (1948) formalized in information theory.

These two models of communication offer radically different lenses through which human language can be seen. On the direct transfer model, communication is at once deterministic (in the sense that words, like physical packages, are assumed to convey a certain determinate content), and singular (in the sense that any given communicative exchange is an isolated event, independent of the broader communicative context, or the prior history of the words or their speakers; Campbell, 1982). By contrast, on the indirect signaling account, communication is predictive and probabilistic. No communicative act occurs in a vacuum; rather, it occurs within the context of a larger linguistic system, governed by extensive, quantifiable regularities. The likelihood of any given message can only be assessed against the distribution of other possible messages that might have been selected instead.

Whereas problems with the first model are well attested (for notable criticisms, see e.g., Wittgenstein, 1953; Quine, 1951; Ramscar & Port, 2015; Baayen & Ramscar, 2015), there is much to recommend the second, particularly as formalized in information theory. Like artificial communication systems, natural languages involve a sender and receiver, a code, and a basic transmission problem. Moreover, they too are indirect means of information exchange (Mandelbrot, 1953; Ramscar & Baayen, 2013). From this perspective, human languages can be seen as complex systems that have evolved over thousands of years and billions of speakers to optimize information flow in communication, and to balance the countervailing demands of learnability and fluent processing (see also Blevins, Milin, & Ramscar, *this volume*).

2.1 *The Discrimination Problem*

> ... consider a coding scheme devised to transmit four experiences: the experience of a fountain, the experience of a fountain pen, the experience of an orange, and the experience of orange juice. Assume a code, shared by encoder and decoder, specifying that the four experiences are signalled by the digit strings 00, 01, 10, and 11, respectively. When seeking to communicate the experience of a fountain pen, the speaker will encode 01, and thanks to the shared code, the listener will decode 01 into the experience of a fountain pen. There is no need whatsoever to consider whether the individual ones and zeroes compositionally contribute to the experiences transmitted. Thus, we can view language [...] as a signal that serves to discriminate complex experiences of the world.
>
> BAAYEN & RAMSCAR (2015)

Within the discriminative framework shared by learning and information theory, language is best described as a probabilistic enterprise in which speakers and listeners cooperate in order to discriminate the content of an intended message from possible alternatives. Formally, the process can be characterized as one of iterative uncertainty reduction: Just as each forking branch in a decision tree further delimits the space of final outcomes, so each utterance (or articulatory gesture) further narrows the range of possible messages (Ramscar & Baayen, 2013). In assessing the dynamics of this process, it is possible to identify both the uncertainty at a given point, and the extent to which it is subsequently reduced. For instance, in context, a speaker's choices can be seen as more or less constrained, corresponding to more or less uncertainty about which word will be uttered next. The more freedom the speaker has in selecting amongst alternatives, the greater the uncertainty, and correspondingly, the more difficult the discrimination problem.

Taking a discriminative approach to communication lays bare the difficulties that nouns pose for language users. In most languages, nouns (both common and proper) are the most *diverse* part of speech, meaning that in any instance in which a noun occurs, the number of other possible alternatives is at its highest, and the discrimination problem is at its peak. This is supported by numerous findings on speech errors. For instance, one of the most common places disfluencies are likely to occur in English is at the determiner preceding a noun; and the more complex the noun is, the more likely a disfluency (Clark & Wasow, 1998). Similarly, nouns are the most common sites for incorrect lexical retrieval and a host of other processing problems (Vigliocco, 1997).

Critically, for our purposes, difficulties such as these have been shown to correlate with the information-theoretic measure of *entropy*, a measure that can be used to quantify the uncertainty over which word will appear in a given context. Entropy offers a particularly useful compression scheme for conceptualizing linguistic uncertainty. While the predictability of a card draw or coin flip is easy to grasp, uncertainty is more difficult to intuit when possible outcomes are numerous, or sequentially dependent, or where probabilities are varied, as is the case for lexical distributions, which comprise thousands of words of widely varying frequencies. Entropy helpfully collapses a multi-dimensional construct down to a single point on a continuum.

Formally, the entropy H over such a distribution is a measure of the expected value of information ('surprisal') over the full range of lexical items (Shannon, 1948):

$$H(p) = H(X) = -\sum_{x \varepsilon X} p(x) log_2 p(x) \qquad (1)$$

When comparing across similarly shaped distributions, entropy will tend to rise as the number of possible outcomes grows. This means that for languages such as English and German, in which the number of noun types outstrips other part-of-speech categories, speakers of both languages will be confronted with most uncertainty when the next item in a sequence is a *noun*. Thus, in example (5), the entropy of possible noun continuations (marked *!*) will be higher than for possible verb continuations (marked #).

(5) Yesterday I # visited the *!* doctor.

Fortunately, speakers have various resources at their disposal for making a particular lexical choice more or less predictable in context. One possibility is to rely on the preceding discourse as a form of scaffolding. For instance, compared to the sparse semantic context provided by (5), the noun *doctor* is far more predictable following the comparatively rich context provided by (6):

(6) Yesterday when I went to the hospital I visited the *! doctor*.

Noun class offers an efficient, systematic way of implementing the same principle. Consider the German equivalent of (5) in (7):

(7)	*Gestern*	*besuchte*	*ich*	*den*	*!Arzt*
	yesterday	visited	I	the.MASCULINE *!*	doctor

While the context is the same as in (5), the uncertainty about the following noun in (7) is greatly reduced by comparison. The following noun must belong to the MASCULINE noun class, and thus nouns of all other genders are eliminated as possible candidates in this context. In short, by systematically partitioning nouns into different classes, a gender marker effectively prunes the space of subsequent possibility, delimiting the set of upcoming nouns to class-consistent possibilities.

There is an accumulating body of evidence that gendered articles guide lexical prediction in precisely this way. Among native speakers of gendered languages, a variety of experimental paradigms, including naming times (Schriefers, 1993), lexical decision (Grosjean et al., 1994), word repetition (Bates et al., 1996), artificial grammar learning (Arnon & Ramscar, 2012), and ERP (Van Berkum et al., 2005; Wicha, Moreno, & Kutas, 2004) have shown that gender facilitates processing when a marker is consistent with a following noun, and inhibits it where there is a mismatch. Auditory gating studies have proved particularly revealing. In such tasks, subjects encounter a word fragment within a clipped auditory sequence, and are asked to produce the target word. When gender information is provided, French subjects correctly identify the target at shorter durations, and with greater confidence. Moreover, gender information not only significantly reduces misidentifications, both in terms of types and tokens, but also limits errors to gender-consistent candidates (Grosjean et al., 1994). In a similar vein, in tip-of-the-tongue (TOT) states, Italian subjects can reliably guess the gender of the noun they are trying to retrieve, even when they cannot produce it (Vigliocco, Antonini, & Garrett, 1997).

These findings are paralleled in studies of visual search. In a study of French speakers, Dahan et al. (2000) asked subjects to view a visual display with a variety of possible referents, while they listened to instructions such as *Cliquez sur le bouton* [*Click on the.MASC button*]. When gender information was provided by the determiner, listeners rapidly shifted their attention to gender-consistent referents, ignoring potential phonological competitors. Lew-Williams and Fernald (2007) report a comparable result for Spanish-speakers, finding that both children and adults are faster to orient to the correct referent on trials when nouns of different genders are displayed than on trials showing nouns of the same gender. Taken together, these results support the conclusion that gendered articles facilitate processing by restricting the space of subsequent possibilities.

2.2 *Managing and Redistributing Entropy*

In understanding the function of gender from this perspective, it is critical to note that gender does not reduce overall entropy, so much as *redistribute* (or manage) it—increasing the entropy of articles, while decreasing the entropy of the nouns that follow them. From a processing perspective, this is consistent with Zipf's famous 'Principle of Least Effort,' which holds that human behavior is shaped by a bias to minimize people's "average rate of work-expenditure over time" (Zipf, 1935; 1949).

On Zipf's loosely psychological account, communicators seek to balance efficiency on the one hand, and comprehensibility on the other, and these opposing forces minimize communicative efforts over time. For example, in a lexicon in which each distinct meaning was assigned a separate word, there would be zero ambiguity, but at a significant processing cost to the speaker engaged in word retrieval. Conversely, a vocabulary comprising a single word would be maximally efficient for the speaker, but "represent the acme of verbal labor" (21) for a listener. Zipf argued that language's characteristic statistical structure reflects a compromise that balances the desire for a many-to-one code (in which there is a single, maximally frequent word) against the desire for one-to-one code (in which there are a vast number of low-frequency words). In terms of optimal coding theory, these balancing forces of unification and diversification can be framed as a compromise between 'word-by-word' coding and 'large-block' coding (Mandelbrot, 1953). Thus, the problem of language design is one of how to distribute the information necessary to discriminate the repertoire of possible messages across acoustic signals (Baayen & Ramscar, 2015).

Once that has been established, the most efficient means of transmitting information across a channel is at a constant rate at (or approaching) the channel's capacity (Shannon, 1948). Indeed, a raft of empirical findings suggest that in accordance with this principle, speakers distribute uncertainty evenly across discourse, in both text and speech. One prediction that comes out of this, is that if the sentences of a given text are equally informative when encountered in context, this is only because the meaning constructed from earlier parts has generated an informative context that reduces the entropy of later parts. In the limit, this suggests that when this contextual scaffolding is stripped away, utterances should become increasingly informative the deeper embedded in the discourse they are. This basic growth pattern has been demonstrated empirically: In a classic study of articles in the Wall Street Journal,

Genzel and Charniak (2002) found that local sentence entropy increases as a function of sentence number, an effect that is driven both by which words are used and how the words are used (i.e., both lexical and syntactic causes). The effect has since been replicated across languages and genres (see also Genzel & Charniak, 2003; Keller, 2004; Qian & Jaeger, 2009).

At the same time, a growing body of evidence supports the idea that in language use, people deftly manage the rate at which information is encoded in linguistic signals, avoiding excessive peaks and troughs in entropy across messages (Aylett & Turk, 2004; Levy, 2008; Jaeger, 2010). One domain in which this has been rigorously tested is speech production, where speakers have been found to smooth information over the acoustic signal by systematically modulating the signal's properties. Varying acoustic duration is one way to accomplish this: articulating unpredictable segments more slowly than predictable ones, and shortening, undershooting, or omitting highly predictable segments (see Gahl, 2012 for a review). These predictions have been substantiated in multiple studies. For instance, Aylett and Turk (2004) found that an inverse relation obtains between a syllable's duration and its predictability in context. Comparable findings on informativity and articulatory effort have been made for words (Bell et al., 2009), morphemes (Pluymaekers, Ernestus, & Baayen, 2005), consonants (Van Son & Van Senten, 2005), and multi-word sequences (Gahl & Garnsey, 2004; Kuperman & Bresnan, 2012). Durational effects have even been replicated in typing (Priva, 2010).

Similarly, in anticipating upcoming words that are information rich, speakers may pause or otherwise delay (Goldman-Eisler, 1958). Predictability also affects specific lexical choices in spontaneous speech and reading aloud: When what they are about to say is predictable, speakers are more likely to employ contractions (Frank & Jaeger, 2008), to omit optional function words (Jaeger, 2010), to use a pronoun referent instead of a full noun-phrase (Tily & Piantadosi, 2009), and to produce fewer disfluencies (Tily et al., 2009). Conversely, when speakers repeat or mimic syntactic constructions in discourse, they temper syntactic redundancy with the selection of more informative, less predictable words (Temperley & Gildea, 2015).

Parallel investigations have been carried out cross-linguistically, with promising results. In a large-scale corpus study spanning eleven Indo-European languages, Piantadosi et al. (2011) found that a word's length is better captured by its average predictability in context than by its raw frequency, with more informative words taking longer forms (see also Manin, 2006). Likewise, in a cross-linguistic comparison of reading aloud data, Pellegrino, Coupé, and Marsico (2011) report that while the various languages under investigation achieve roughly comparable rates of information transfer overall, they strike markedly different balances between information density and speech rate in doing so: in languages with less information per syllable, syllables tend to be spoken faster, and vice versa.

These findings make clear that language distributions (and speakers) ably regulate the uncertainty associated with temporal dynamics and lexical choices. Gender markers may simply serve as another resource by which to accomplish this: If gendered articles serve to redistribute nominal entropy, this will smooth potential spikes in information, helping speakers maintain a more constant entropy rate.

3 Noun Class and Entropy Reduction in German

Our proposal is that noun class systematically narrows the set of candidates that follow a gender marker, thereby reducing the amount of information that a noun conveys on its own. As a first test of this hypothesis, we conducted an analysis of nominal entropy distributions in the German mega-corpus Stuttgart deWaC.[3] German is a language with a binary number system (singular and plural), three-class gender system (masculine, feminine, and neuter), and four grammatical cases in which nouns can occur (nominative, accusative, dative, and genitive). Accordingly, to assess the influence of gender marking on nominal entropy, the entropy of all the nouns within each case (2) was compared to the conditional entropy of those nouns following articles marked for gender and number (3).

$$
\begin{aligned}
H(N) &= -\sum_i P(N_i) log_2 P(N_i) \\
&\approx -\sum_i \frac{\text{Count}(N_i)}{\text{Total}(N)} log_2 \frac{\text{Count}(N_i)}{\text{Total}(N)} \qquad (2)
\end{aligned}
$$

3 The SdeWaC is a subset of the WaCky corpus, which comprises more than 44 M sentences, 850M word tokens, and 1.1 M word types (Faaß & Eckart, 2013; Baroni, Bernardini, Ferraresi, & Zanchetta, 2009). The corpus was first annotated with fine-grained part-of-speech categories using the RFTagger (Schmid & Laws, 2008), and article contractions were expanded (*im* -> *in dem*). Every noun that immediately followed a definite article was extracted with its gender, case, and number tags, and tabulated.

For instance, for nouns following the masculine nominative article *der*, the conditional entropy would be given by:

$$H(N) = -\sum_i P(N_i \mid der) log_2 P(N_i \mid der)$$
$$\approx -\sum_i \frac{\text{Count}(N_i \mid der)}{\text{Total}(N \mid der)} log_2 \frac{\text{Count}(N_i \mid der)}{\text{Total}(N \mid der)} \quad (3)$$

Consistent with our suggestion that German gender serves to reduce uncertainty about upcoming nouns in discourse, we found:

Entropy (nouns) > Entropy (nouns|definite article)
> Entropy (nouns|definite article, case)
> Entropy (nouns|definite article, case, gender)

These results show that, as expected, in each of the German cases, gender markers significantly reduce nominal entropy (Figure 9.1; The same qualitative results were obtained in an analysis of the Negra II corpus of German newspapers, Skut et al. 1997).

To further test this hypothesis, we then examined the effect of noun class marking on the distribution of nouns in German. By effectively partitioning the noun space, gender markers should offload some of the uncertainty about the upcoming noun onto the determiner, thereby smoothing entropy over the marker-noun pairing. Accordingly, when prenominal class marking is present, the following noun should be relatively well-predicted, compared to cases in which its class goes unmarked. Assuming that communicators aim to keep uncertainty relatively constant, and that gender marking offers an effective means of selectively modulating uncertainty, German speakers should make use of a greater variety of nouns when noun class marking is present than when it is absent.

The German plural offers an illustrative test case. While all German singular nouns are marked for gender, plural nouns are not. Accordingly, following a definite article, speakers should employ a more diverse (and more informative) set of nouns in the singular than in the plural. A measure of the difference in the overall lexical diversity of the two noun types in this context can be estimated by calculating their type/token ratio (while holding the sample size constant), with a higher type/token ratio suggesting a greater diversity of nominal usage. Conveniently, this metric is simply the inverse of average frequency, allowing for a straightforward test of this hypothesis: the lower the average frequency, the greater the diversity of nominal usage.

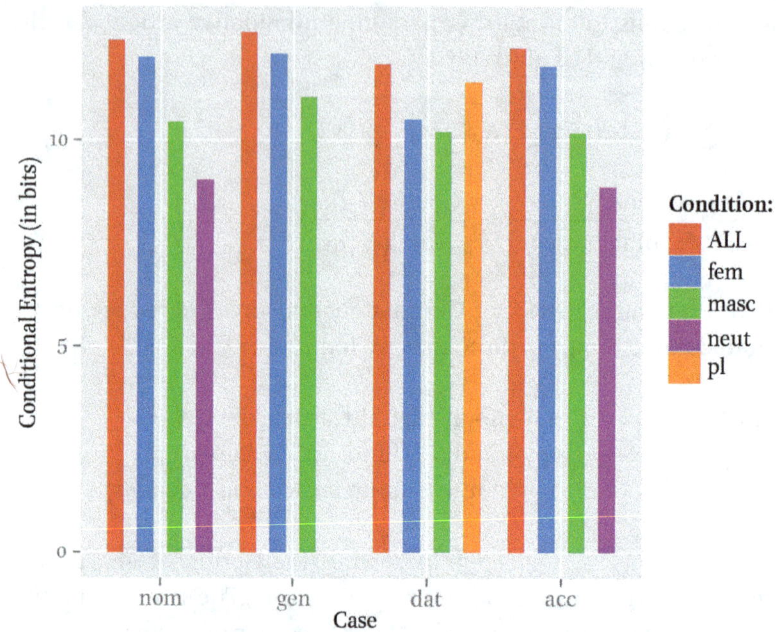

FIGURE 9.1 *Noun entropy conditioned on case and number, irrespective of gender vs. gender sensitive. Notice that because of syncretism, not every category is represented independently for each case; German lacks any morphological distinction between feminine and plural articles in the nominative, accusative, and genitive cases, and between masculine and neuter articles in the dative and genitive cases. In this analysis, forms that took the same marker within a given case were tabulated together (e.g., for nominative, both feminine and plural nouns contribute to the entropy calculation for 'die').*

Consistent with our hypothesis, an examination of Determiner-Noun contexts in the SdeWaC revealed that the singular nouns in our sample had a higher type/token ratio than the plural nouns. When the frequencies for singular and plural nouns are normalized to per-million occurrences, the mean frequency for singular nouns is 0.75, and that of plurals is 1.43; correspondingly, lexical diversity for singular and plural nouns is 1.33 and 0.70, respectively. German plurals, which are not gender-marked, thus show a substantial reduction in lexical diversity relative to singulars, indicating that gender catalyzes the use of a wider array of nominal forms.

Figure 9.2 shows the distributional impact of gender-marking. The figure represents the (extremely) skewed frequency distributions for singular (gender-marked) and plural (unmarked) nouns following determiners. The Lotka-plot

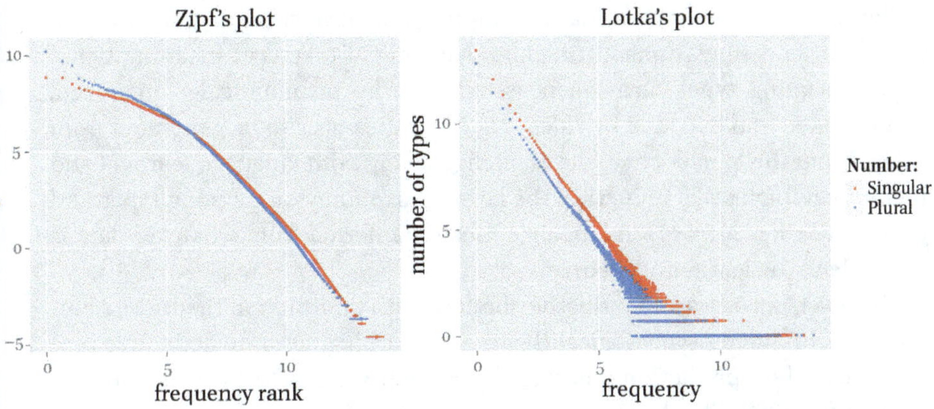

FIGURE 9.2 *The frequency distributions of German singular and plural nouns following a determiner. These distributions are plotted in two complementary ways: While the Zipf-plot (left panel) plots frequency rank by frequency, the Lotka-plot (right panel) plots frequency by number of different word types; both are shown in a log-log plane. In a sense, the plots are showing each other's tails (c.f., Chen & Leimkuhler, 1986; Kunz, 1987).*

indicates that whereas there are significantly more singular noun types with low frequencies, the inverse is true for plurals, which cover a wider range of the most frequent types. The Zipf-plot echoes this trend, revealing that the difference in nominal frequencies is most pronounced in the high frequency range; within that uppermost band, a singular noun of a given frequency rank will (on average) be markedly lower in frequency than its plural equivalent.

Interestingly, when determiners are treated as mere case markers, independent of gender and number, and their following distributions are analyzed separately, the lexical diversity of following nouns is *equal*, on average. This is, again, consistent with the suggestion that languages (and hence speakers) are finely attuned to the uncertainty of their productions, exploiting the varied resources at their disposal to keep entropy smoothed.

3.1 *Semantics and the Function of Noun Class*

One question that arises is whether this partitioning of nouns into classes is arbitrary, or whether there might be a hidden logic behind it. Recall the damning words of Mark Twain—why the sexless young maiden, the female tomcat? Is there really no sense or sensibility to gender? In order to understand how noun class might best be configured to facilitate communication, it is important to consider the wider functional implications of uncertainty reduction in language use.

Recall that under the standard metaphor, language is conceptualized as a process of encoding, transmitting and decoding of tokens of meaning types. These meaning types have been assumed to be taxonomically organized, and encoded and decoded by rules that allow messages to be generated from them. From this perspective, the challenge facing both language learners and theoretical linguists is inductive: the correct taxonomy of meaning types and generative rules for a given language must be inferred from whatever data is available to the learner or theorist.

However, in contrast to inductive models, both information theory and empirically grounded psychological theories of learning describe deductive processes based on prediction and discrimination (Ramscar & Baayen, 2013; see Shannon, 1948; Kullback & Leibler, 1951; Rescorla & Wagner, 1972). Information theory sees "the fundamental problem of communication [as] that of reproducing at one point, either exactly or approximately, a message selected at another point" (Shannon, 1948). Seen from this perspective, communication need not consist in the transmission of tokens corresponding to fixed semantic types between speakers. Rather, it can be seen as a process in which a speaker reduces a hearer's uncertainty about the meaning of a message by whatever means are available (Ramscar et al, 2010).

For example, when a German speaker uses the expression *der Hund* "the dog", the masculine-gendered article *der* helps the hearer to expect a dog as the referent. From this perspective, not only the token *Hund* but also the gendered article *der*—and indeed the entirety of the surrounding predictive context, including any verbs and adjectives—helps the hearer to form the belief that the speaker wants to say something about a "dog". Linguistic communication can thus be seen as a probabilistic process in which a speaker helps a listener to predict, either exactly or (more often) approximately, the speaker's intentions.

These very different models of the way that language works yield very different predictions about the function of noun class. The taxonomic approach leads naturally to the prediction that noun class adds to the taxonomy of meaningful words in a given language. For example, Lakoff's (1986) 'Domain of Experience Principle' holds that nouns that occur in similar contexts tend to have the same gender. Gender is seen as mapping to an abstract "semantic field" for the purposes of transmitting meaning. However, since the task of identifying exactly what these semantic fields actually *are* has proven to be difficult in many languages, this idea is often augmented by the supplementary assumption that although many semantic fields have a dominant gender, each of these comes with a set of individually specified exceptions (Zubin & Köpcke 2009, 1981; Zubin 1992). That is, from the taxonomic perspective, noun classes tend to map to a semantic field, except in all cases where they *don't*.

However, a system in which semantic regularities are not immediately straightforward might actually *benefit* discrimination. The benefits of such dispersion are noted by Zubin and Köpcke (1986), who show that while nouns in the semantic class of 'kitchen implements' in German are evenly distributed among the genders without any obvious or sensible pattern, the "patterning" of their dispersal may actually facilitate reference tracking among objects. In short, the suggestion is that when nouns that occur in similar contexts are assigned different genders, this facilitates discrimination between possible referents.

On the other hand, Twain may have been exaggerating slightly; there do appear to be semantic patterns to gender assignment in German, though they are rife with exceptions (Zubin & Köpcke, 1986; Lakoff, 1986). The existence of such semantic regularities makes clear that noun classes are not distributed so that *all* similar nouns receive different genders. (In the limit, such a scheme might prove redundant from the point of view of maintaining a constant entropy rate.) Instead, the German gender system may be optimized in a different way. For example, almost all German alcoholic drinks are in the masculine class, except *beer*, which is neuter. This state of affairs is mirrored for non-alcoholic beverages, which also tend to be in the masculine class, with one notable exception—*water*, which is also neuter (as are the common words for drink and beverage). Taxonomically, a grouping of gin and juice and coffee on the one hand, and beer and water on the other, makes little sense. Yet a class division between the drinks that might be more or less expected in a given context does, because a deductive discriminative process works by eliminating possible interpretations that are *not* intended (Shannon, 1948; Rescorla & Wagner, 1972; Ramscar et al, 2010).

To flesh out this idea, compare the information requirements for helping someone predict that Beethoven rather than Mozart will be the topic of a sentence, as compared to helping someone predict that it will be Villa Lobos rather than Schoenberg (lesser known 20th Century composers). In discourse about composers of classical music, both Beethoven and Mozart—by dint of their fame and presence in any educated Westerner's general knowledge—will be highly predictable in context. Thus, while much could be gained from deploying a contextual cue that eliminates Mozart as a possible topic as opposed to Beethoven (or vice versa), little could be gained from eliminating the relatively obscure and unpredictable Villa Lobos or Schoenberg. Contextual cues that are specifically informative about high frequency items will be very useful for discriminating between those items.

Even if the topic of discourse does turn out to be Villa Lobos or Schoenberg, cues that favor the elimination of highly predictable competitors will still be more valuable than cues that favor the specific prediction of one over the other. Because Mozart and Beethoven will be strongly expected candidates in discourse about composers, a cue that eliminated one or both of them from consideration would be a boon for communicative clarity, as it would improve the predictability of *both* Villa Lobos and Schoenberg. This is not to say that contextual information that discriminates Villa Lobos from Schoenberg might not also be helpful here, but rather that that information will only be relevant *after* competition from Mozart and Beethoven has been reduced or eliminated.

As this example illustrates, depending on the distribution of items in a semantic class, both semantic clustering and semantic dispersal could be employed to optimize the use of gender information for discriminating between alternatives of differing probabilities. For example, to assist with overall entropy reduction, a noun class system might fruitfully assign Beethoven and Mozart to their own classes, while grouping Villa Lobos and Schoenberg together in another. Indeed, in terms of informativity, it might be perfectly sensible if Villa Lobos and Schoenberg were classed alongside more obscure composers from other classical periods, even if this makes relatively little sense taxonomically. This logic can begin to help explain why German puts what are historically its most common drinks—beer and water—in a class apart from most other beverages.

3.2 *Testing Semantics*

The notion that German noun class is informative is compatible with both a taxonomic and a discriminatory approach to language. To the extent that the two approaches make different predictions, the differences are in the details. While both models predict a correlation between semantics and gender, the taxonomic model leaves the exact nature of that relation opaque and filled with exceptions. By contrast, the discriminatory model suggests that 'exceptions' are likely nothing of the sort, reflecting instead the properties of the underlying system. The discriminatory model thus makes an intriguing prediction: not only should we expect to find a positive correlation between semantics and noun class, but we should also be able to detect systematic patterns where semantics and noun class diverge.

For a gender system to be maximally functional, it needs to reduce the uncertainty of an upcoming noun in context by *narrowing the search space* of likely candidates. That is, it needs to discriminate against alternative nouns on the basis of their likelihood. To ideally meet this requirement, such a system should assign different genders to nouns that are both semantically similar and potentially confusable in context. This raises an intriguing question: how might this be achieved? In practice, semantic considerations at the local and discourse level will have significantly altered the shape of likelihood distribution, making some nouns far more likely in context, and others considerably less so. However, absent a means of making entropy reduction semantically interpretable—a task that is beyond the scope of the current work—this observation is less than illuminating.

To try and shed some light on this question, consider the following possibilities:

1) Noun class might discriminate semantically similar nouns that co-occur together regularly, such as gin versus tonic, or coffee versus tea.
2) Noun class might discriminate semantically similar nouns that differ by frequency, such as water (high-frequency) versus root beer (low-frequency).
3) Noun class might discriminate nouns that are highly likely in a certain context, but which are semantically distinct, such as drinks versus food.

As we noted in our example above, the degree to which one strategy or another is most appropriate for a given noun will depend both upon its overall likelihood, and the degree to which it is already predicted when a gender marker occurs. Thus 1) will work better when there is a higher degree of certainty about the specific noun that will occur, whereas 3) will be a better fit when there is a lower degree of certainty. The degree of uncertainty will always depend on the specifics of the particular noun: its frequency, the frequency of its neighbors, and the contexts in which it (and its neighbors) are encountered. An optimal system should tailor its level of support for each noun based on these factors, supplying different information depending both upon the overall likelihood of the noun, and the likelihood of there being other discriminatory information available in context.

To gain a better understanding of how this might function in German, we examined the fine-grained relationship between semantics, contextual confusability, and noun class. Specifically, using a Generalized Additive Model (GAM) with binomial link-function (*mgcv* package in R Statistical Computing Environment; see Wood, 2006; 2011; R Core Team, 2015), we attempted to predict gender sameness for pairs of nouns based on the pair's frequency, pointwise mutual information (PMI), and semantic similarity.[4] Our modeling results

4 To gather the necessary input to the model, the RFTagger was first run over the SdeWaC, expanding article contractions and lemmatizing noun forms (Schmid & Laws, 2008). Nouns that occur in the corpus in all case and number permutations were then selected for analysis (61K in total), with individual frequency tabulated as a lemma count, and co-occurrence

were validated with a bootstrap sampling technique (N=1000 simulation runs). To remain conservative with respect to Type-I errors, we report the most likely values for the test statistic and the maximal p-values across all runs.

The model revealed that among noun pairs, overall gender sameness was predicted by two composite factors: (1) the frequency of the words in the pairing; and (2) the semantic similarity of the pair modulated by their co-occurrence likelihood (Figure 9.3). In the case of (1), the model indicated that the lower the word frequencies of the pair, the more likely they were to belong to the same noun class ($\chi2 = 599.38$; $p < 0.0001$). In the case of (2), it was found that the more tightly semantically coupled a pair of nouns, the more likely they were to share gender. However, this pattern was further modulated by the

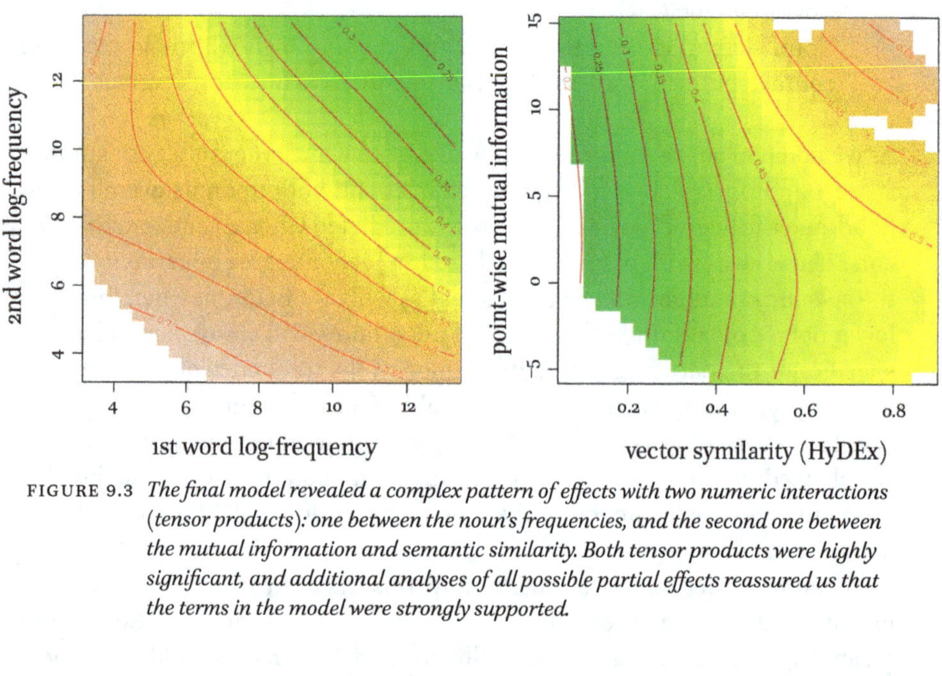

FIGURE 9.3 *The final model revealed a complex pattern of effects with two numeric interactions (tensor products): one between the noun's frequencies, and the second one between the mutual information and semantic similarity. Both tensor products were highly significant, and additional analyses of all possible partial effects reassured us that the terms in the model were strongly supported.*

---

rates between noun pairs calculated within a 2-word bidirectional window. These frequency and co-occurrence counts were used to compute PMI, a measure of association that compares the probability of two nouns co-occurring against the probability of them occurring independently (Church & Hanks, 1989). Finally, the semantic similarity of noun pairs was calculated by running the High Dimensional Explorer (HiDEx; Shaoul & Westbury, 2010) over the lemmatized corpus, using a 5-word bidirectional window, and inverse linear ramp weighting. HiDEx is an implementation of the hyperspace analog to language (HAL) semantic space model, which stores raw lexical co-occurrence information in a high-dimensional matrix that it subjects to a series of transforms, yielding semantic similarity relations.

mutual information between the noun pair ($\chi2 = 711.43$; $p < 0.0001$): When the likelihood that the two nouns systematically co-occurred together was low, the effect of semantic similarity was attenuated; conversely, as co-occurrence likelihood increased, the effect of semantic similarity grew stronger. Thus, a noun pair was most likely to share the same gender when its nouns were both highly informative of one another and also contextually very similar.

These results provide comprehensive quantitative support for the idea that there are systematic semantic trends in noun class assignment, indicating that while nouns that are semantically similar tend to belong to the same gender, this effect is modulated by frequency. Whereas high-frequency items tend to be distributed *across* genders, low-frequency items tend to be clustered within the *same* gender. Hence, the gender marking system in German appears to make use of both semantic clustering and semantic dispersion, with the choice of strategy varying with frequency.

An additional question worth pursuing is whether these strategies are realized differently in different classes. In fact, the probabilities of nominal gender in German differ markedly, with nearly half of nouns classed as feminine (49.45%), roughly a third as masculine (31.64%), and close to a fifth as neuter (18.96%). To assess whether nouns might pattern into different genders on the basis of their frequency, we attempted to predict noun class from noun frequency, using Bayesian multinomial logistic regression (*BayesLogit* package in R Statistical Computing Environment; see: Polson, Scott, & Windle, 2013; R Core Team, 2015).[5] This analysis revealed that noun frequency does *not* predict noun class (Figure 9.4). Thus, while there appear to be strong general biases in class assignment, these biases do not pattern by frequency, suggesting that the different classes likely share quite similar distributional properties.

4 Why Taxonomy Misses the Point

The studies reported here provide evidence in support of our suggestion that German noun class is well-adapted for the purpose of helping communicators predict nouns in context. The dispersal of nouns across different gender classes is clearly sensitive to factors that influence an item's discriminability,

5 To make the computation feasible, the algorithm was run over a randomly selected sample of 12,000 nouns. Markov Chain Monte Carlo (MCMC) sampling was applied to obtain the posterior distribution of the regression parameters. The first 1,000 iterations were excluded as part of an initial burn-in, after which results were analyzed for 10,000 MCMC iterations.

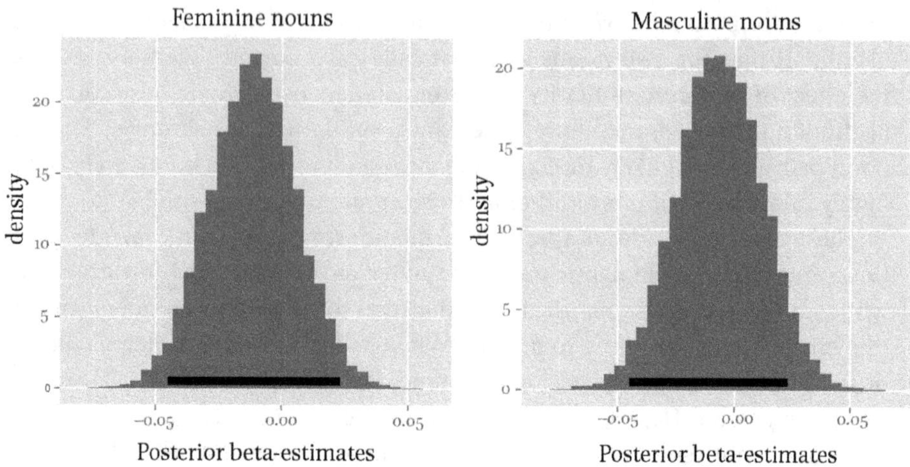

FIGURE 9.4 *The posterior densities of the estimated coefficients for the frequency covariate for feminine (left panel) and masculine nouns (right panel). Bayesian credible intervals (95% HPD) are marked with a black horizontal line. As can be seen, the coefficients fall close to zero, and range over both positive and negative values. Such a result indicates that masculine and feminine nouns are distributionally indistinguishable from neuter nouns (the reference level in the model).*

and appears structured to *level* the effects of these factors, making nouns more equally predictable in context.

While it has often been claimed that the German gender system is unsystematic and meaningless, our findings suggest, to the contrary, that not only does noun class serve to efficiently manage nominal entropy, but also that—like many other subsystems of language—gender in German is more specifically informative about high-frequency nouns than low-frequency nouns. It is notable that verb inflection, in both German and English, shares the same pattern: high-frequency verbs tend to have specific (irregular) inflection patterns that are highly informative about the inflected form of a given verb, whereas low-frequency verbs have generic (regular) inflection patterns that are less specifically informative (Baayen & Moscoso del Prado Martín, 2005).

While this point may seem counterintuitive, it is well predicted by a discriminative account. High and low-frequency forms pose markedly different challenges in terms of entropy management. Compared to lower frequency forms, high frequency items tend to be more contextually 'promiscuous' (Adelman, Brown, & Queseda, 2006), to be more semantically similar to more other words (Steyvers & Tenenbaum, 2005), and to have denser phonological neighborhoods (Andrews, 1992), meaning that they are, at once, less disambiguated by context, and more confusable with other items. At the same time, high-frequency items are more likely to be encountered in sparser, less distinctive linguistic contexts, where their prediction is unsupported by other material (Genzel & Charniak, 2002, 2003; Sigurd, Eeg-Olofsson & van de Weijer, 2004). Accordingly, since context will often fail to distinguish highly frequent (and thus highly likely) nouns, a great deal of uncertainty is inevitable, and a greater level of uncertainty reduction is called for.

The relation between gender and frequency also makes sense in terms of learnability: rigid, highly informative conventions, such as gender marking, can only arise in a language if all of the speakers in a community reliably encounter and acquire them. In German, the distribution of nouns will support the learning of apparently 'arbitrary' gender markers for more common nouns, because by dint of their frequency in the input, these nouns will be encountered by young learners early in development, at a sensitive period in cortical maturation (Ramscar & Gitcho, 2007; Thompson-Schill, Ramscar, & Chrysikou 2009). Accordingly, the learning of these forms will not be influenced by the top-down factors that inhibit the acquisition of irregulars in adults (Ramscar, Dye, & McCauley, 2013). By contrast, the rarity (or complete absence) of low-frequency nouns in child-directed speech will render the rote learning of their gender classes all but impossible (Blevins, Milin & Ramscar, *this volume*). Instead, the presence of converging semantic and acoustic cues will serve to make the gender of low-frequency nouns predictable (i.e., 'regular'; Frigo & McDonald, 1998). This neatly solves the problem of how to mark nouns in a system that, because of its highly skewed distribution, renders the task of learning new noun-forms and their classes a continuous process that is never complete—leaving the system supple and adaptable to the demands of learning across the lifespan (Ramscar, Hendrix, Love & Baayen, 2013; Ramscar, Hendrix, Shaoul, Milin & Baayen, 2014).

All of which is to say that the structure of German noun class is shaped by considerations that are the *opposite* of those that have traditionally been understood to determine gender assignment. Noun class serves a discriminatory purpose, and the information processing requirements of discriminatory and taxonomic systems are very different. Masculine and feminine gender classes do not reflect any kind of deep underlying taxonomic distinction; rather, items are assigned to different gender classes because assigning them to different gender classes is systematically informative. While gender classification can appear to be taxonomically interpretable at a squint, up close, many of its classifications appear taxonomically senseless. Yet there is an underlying logic to the system that is evident throughout: gender serves to redistribute the entropy of nouns, making them more predictable, on average, in context. When we at last dispense with the long-standing assumption that gender marking is taxonomic, we dispense too with the confusions that have plagued its study.

References

Adelman, J.S., Brown, G.D.A., & Quesada, J.F. 2006. Contextual diversity, not word frequency, determines word-naming and lexical decision times. *Psychological Science* 17(9). 814–823.

Andrews, S. 1992. Frequency and neighborhood effects on lexical access: Lexical similarity or orthographic redundancy? *Journal of Experimental Psychology: Learning, Memory, and Cognition* 18(2). 234–254.

Arnon, I. & Ramscar, M. 2012. Granularity and the acquisition of grammatical gender: How order of acquisition affects what gets learned. *Cognition* 122(3). 292–305.

Aylett, M., & Turk, A. 2004. The smooth signal redundancy hypothesis: A functional explanation for relationships between redundancy, prosodic prominence, and duration in spontaneous speech. *Language and Speech* 47(1). 31–56.

Baayen, R.H., Milin, P., Filipovic Durdevic, D., Hendrix, P., & Marelli, M. 2011. An amorphous model for morphological processing in visual comprehension based on naive discriminative learning. *Psychological Review* 118(3). 438–481.

Baayen, R.H. & Ramscar, M. 2015. Abstraction, storage, and naive discriminative learning. In Dabrowska, E. and D. Divjak (eds.), *Handbook of Cognitive Linguistics*, 99–120. De Gruyter Mouton.

Baroni, M. Bernardini, S., Ferraresi, A., & Zanchetta. E. 2009. The WaCky wide web: A collection of very large linguistically processed web-crawled corpora. *Language Resources and Evaluation* 43(3). 209–226.

Bates, E., Devescovi, A., Hernandez, A., & Pizzamiglio, L. 1996. Gender priming in Italian. *Perception & Psychophysics* 58(7). 992–1004.

Bell, A., Brenier, J.M., Gregory, M., Girand, C., & Jurafsky, D. 2008. Predictability effects on durations of content and function words in conversational English. *Journal of Memory and Language* 60(1). 92–111.

Blevins, J., Milin, P. & Ramscar, M. (*this volume*) Zipfian discrimination.

Bloomfield, L. 1933. *Language*. New York: Holt.

Campbell, J. 1982. *Grammatical man: Information, entropy, language and life*. New York: Simon & Schuster.

Chen, Y.S., & Leimkuhler, F.F. 1986. A relationship between Lotka's law, Bradford's law, and Zipf's law. *Journal of the American Society for Information Science* 37(5). 307–314.

Church, K.W., & Hanks, P. 1989. Word association norms, mutual information, and lexicography. *Proceedings of the 27th Annual Meeting on Association for Computational Linguistics*, Morristown, NJ, USA: Association for Computational Linguistics. 76–83.

Clark, H.H., & Wasow, T. 1998. Repeating words in spontaneous speech. *Cognitive Psychology* 37. 201–242

Claudi, U. 1985. *Zur Entstehung von Genussystemen: Überlegungen zu einigen theoretischen Aspekten, verbunden mit einer Fallstudie des Zande*. Hamburg: Buske.

Corbett, G.G. 1991. *Gender*. Melbourne: Cambridge University Press.

Dahan, D., Swingley, D., Tanenhaus, M.K., & Magnuson, J.S. 2000. Linguistic gender and spoken-word recognition in French. *Journal of Memory and Language* 42(4). 465–480.

Faaß, G., & Eckart, K. 2013. SdeWaC—A corpus of parsable sentences from the web. Gurevych, Iryna, Chris Biemann & Torsten Zesch (eds.): *GSCL 2013, LNCS 8105*. Heidelberg: Springer.

Frank, A.F., & Jaeger, T.F. 2008. Speaking rationally: Uniform information density as an optimal strategy for language production. *Proceedings of the 30th Annual Meeting of the Cognitive Science Society*.

Frigo, L., & McDonald, J.L. 1998. Properties of phonological markers that affect the acquisition of gender-like subclasses. *Journal of Memory and Language* 39. 218–245.

Gahl, S., & Garnsey, S. 2006. Knowledge of grammar includes knowledge of syntactic probabilities. *Language* 82(2). 405–410.

Gahl, S., Yao, Y., & Johnson, K. 2012. Why reduce? Phonological neighborhood density and phonetic reduction in spontaneous speech. *Journal of Memory and Language* 66(4). 1–18.

Genzel, D., & Charniak, E. 2002. Entropy rate constancy in text. In *Proceedings of the 40th Annual Meeting on Association for Computational Linguistics*. Association for Computational Linguistics: Morristown, NJ. 199–206.

Genzel, D., & Charniak, E. 2003. Variation of entropy and parse tree of sentences as a function of the sentence number. In *Proceedings of the Conference on Empirical Methods in Natural Language Processing*: Sapporo, Japan. 65–72.

Goldman-Eisler, F. 1958. Speech production and the predictability of words in context. *The Quarterly Journal of Experimental Psychology* 10(2). 96–106.

Grosjean, F., Dommergues, J.Y., Cornu, E., Guillelmon, D., & Besson, C. 1994. The gender-marking effect in spoken word recognition. *Perception & Psychophysics* 56(5). 590–598.

Hahn, U., & Nakisa, R.C. 2000. German inflection: Single route or dual route? *Cognitive Psychology* 41(4). 313–360

Heath, J. 1975. Some functional relationships in grammar. *Language* 51. 89–104.

Holzmann, G.J., & Pehrson, B. 1994. *The early history of data networks*. Wiley-IEEE Computer Society Press.

Jaeger, T.F. 2010. Redundancy and reduction: Speakers manage syntactic information density. *Cognitive Psychology* 61(1). 23–62.

Keller, F. 2009. The entropy rate principle as a predictor of processing effort: an evaluation against eye-tracking data. In *Proceedings of the Conference on Empirical Methods in Natural Language Processing*, Barcelona, Spain. 317–324.

Kilarski, M. 2007. On grammatical gender as an arbitrary and redundant category. In Kilbee, D. (ed.), *History of Linguistics 2005: Selected papers from the 10th International*

Conference on the History of Language Sciences (ICHOLS X). John Benjamins, Amsterdam. 24–36.

Köpcke, Klaus-Michael. 1982. *Untersuchungen zum Genussystem der deutschen Gegenwartssprache* (Linguistische Arbeiten 122). Tübingen: Niemeyer.

Koval', A.I. 1979. O značenii morfologičeskogo pokazatelja klassa v fula. In N.V. Oxotina (ed.) *Morfonologija i morfologija klassov slov v jazykax Afriki*. Moscow: Nauka. 5–100.

Kullback, S. & Leibler, R.A. 1951. On information and sufficiency. *Annals of Mathematical Statistics* 22 (1). 79–86.

Kuperman, V., & Bresnan, J. 2012. The effects of construction probability on word durations during spontaneous incremental sentence production. *Journal of Memory and Language* 66(4). 1–24.

Kunz, M. 1988. Lotka and Zipf: Paper dragons with fuzzy tails. *Scientometrics* 13(5–6). 289–297.

Lakoff, George. 1986. Classifiers as a reflection of mind. In Craig, C. (ed.), *Typological Studies in Language 7: Noun Classes and Categorization*, 13–51.

Levy, R. 2008. Expectation-based syntactic comprehension. *Cognition* 106(3). 1126–1177.

Lew-Williams, C., & Fernald, A. 2007. Young children learning Spanish make rapid use of grammatical gender in spoken word recognition. *Psychological Science*, 18(3), 193–198.

MacWhinney, B., Bates, E., & Kliegl, R. 1984. Cue validity and sentence interpretation in English, German, and Italian. *Journal of Verbal Learning and Verbal Behavior* 23(2). 127–150.

Mandelbrot, B. 1953. An informational theory of the statistical structure of language. *Communication Theory* 84. 486–502.

Manin, D. 2006. Experiments on predictability of word in context and information rate in natural language. *Journal of Information Processes* 6. 229–236.

Maratsos, M.P. 1979. How to get from words to sentences. In D. Aaronson & R. Rieber (eds.), *Perspectives in Psycholinguistics*. Hillsdale, N.J.: Erlbaum.

Miozzo, M., & Caramazza, A. 1999. The selection of determiners in noun phrase production. *Journal of Experimental Psychology: Learning, Memory, and Cognition* 25. 907–922.

Pellegrino, F., Coupé, C., & Marsico, E. 2011. A cross-language perspective on speech information rate. *Language* 87(3). 539–558.

Piantadosi, S.T., Tily, H., & Gibson, E. 2011. Word lengths are optimized for efficient communication. *Proceedings of the National Academy of Sciences* 108(9). 3526–9.

Pluymaekers, M., Ernestus, M. and Baayen, R.H. 2005. Articulatory planning is continuous and sensitive to informational redundancy. *Phonetica* 62. 146–159.

Polson, N.G., Scott, J.G., & Windle, J. 2013. Bayesian inference for logistic models using Pólya-Gamma latent variables. *Journal of the American Statistical Association* 108(504). 1339–1349.

Priva, U.C. 2010. Constructing typing-time corpora: A new way to answer old questions. *Proceedings of the 32nd Annual Meeting of the Cognitive Science Society*.

Qian, T., & Jaeger, T.F. 2012. Cue effectiveness in communicatively efficient discourse production. *Cognitive Science* 36(7). 1312–1336.

Quine, W.V.O. 1951. Two dogmas of empiricism. *The Philosophical Review* 60. 20–43.

Ramscar, M. & Baayen, H. 2013. Production, comprehension and synthesis: A communicative perspective on language. *Frontiers in Language Sciences* 4. 233.

Ramscar, M., Dye, M. & McCauley, S. 2013. Error and expectation in language learning: The curious absence of 'mouses' in adult speech. *Language* 89(4). 760–793.

Ramscar, M., Hendrix, P., Love, B. & Baayen, H. 2013. Learning is not decline: The mental lexicon as a window into cognition across the lifespan. *The Mental Lexicon* 8(3). 450–481.

Ramscar, M., Hendrix, P., Shaoul, C., Milin, P. & Baayen, R.H. 2014. The myth of cognitive decline: Non-linear dynamics of lifelong learning. *Topics in Cognitive Science* 6. 5–42.

Ramscar, M. & Port, R. 2015. Categorization (without categories). In E. Dawbroska & D. Divjak (Eds.), *Handbook of Cognitive Linguistics*. De Gruyter Mouton.

Ramscar, M., Yarlett, D., Dye, M., Denny, K., & Thorpe, K. 2010. The effects of feature-label-order and their implications for symbolic learning. *Cognitive Science* 34. 909–957.

Ramscar, M., & Gitcho, N. (2007). Developmental change and the nature of learning in childhood. Trends in Cognitive Sciences, 11(7), 274–279.

Reddy, M.J. 1979. The conduit metaphor: A case of frame conflict in our language about language. In *Metaphor and Thought*.

Rescorla, R.A., & Allan R. Wagner, A.R. 1972. A theory of Pavlovian conditioning: Variations in the effectiveness of reinforcement and nonreinforcement. In Black, A.H., and Prokasy, W.F. (Eds.), *Classical conditioning II: Current research and theory* (pp. 64–99). New York: Appleton-Century-Croft.

Schmid, H. & Laws, F. 2008. Estimation of conditional probabilities with decision trees and an application to fine-grained POS tagging. *COLING 2008*, Manchester, Great Britain.

Schriefers, H. 1993. Syntactic processes in the production of noun phrases. *Journal of Experimental Psychology: Learning, Memory, and Cognition* 19. 841–850.

Schriefers, H., & Teruel, E. 2000. Grammatical gender in noun phrase production: the gender interference effect in German. *Journal of Experimental Psychology: Learning, Memory, and Cognition* 26(6). 1368–1377.

Shannon, C.E. 1948. A mathematical theory of communication. *Bell System Technical Journal* 27. 379–423, 623–656.

Shaoul, C. & Westbury, C. 2010. Exploring lexical co-occurrence space using HiDEx. *Behavior Research Methods* 42(2). 393–413.

Sigurd, B., Eeg-Olofsson, M., van de Weijer J. 2004. Word length, sentence length and frequency—Zipf revisited. *Studia Linguistica* 58. 37–52.

Skut, W., B. Krenn, T. Brants., & H. Uszkoreit. 1997. An annotation scheme for free word order languages. In *Proceedings of the Fifth Conference on Applied Natural Language Processing* (ANLP), Morgan Kaufmann, San Francisco.

Sperber, D. & Wilson, D. 1996. *Relevance: Communication and cognition.* (2nd ed.) Wiley-Blackwell.

Steyvers, M., & Tenenbaum, J.B. 2005. The large-scale structure of semantic networks: statistical analyses and a model of semantic growth. *Cognitive Science* 29(1). 41–78.

Team, R.C. 2015. R: A language and environment for statistical computing. Vienna, Austria; 2014. *URL http://www.R-project.org.*

Temperley, D., & Gildea, D. 2015. Information density and syntactic repetition. *Cognitive Science.* doi: 10.1111/cogs.12215.

Thompson-Schill, S., Ramscar, M., & Chrysikou, E. 2009. Cognition without control: when a little frontal lobe goes a long way. *Current Directions in Psychological Science* 8(5). 259–263.

Tily, H., & Piantadosi, S. 2009. Refer efficiently: Use less informative expressions for more predictable meanings. In *Proceedings of the Workshop on the Production of Referring Expressions: Bridging the Gap between Computational and Empirical Approaches to Reference.*

Tily, H., Gahl, S., Arnon, I., Snider, N., Kothari, A., & Bresnan, J. 2009. Syntactic probabilities affect pronunciation variation in spontaneous speech. *Language and Cognition* 1(2). 147–165.

Van Berkum, J.J.A., Brown, C.M., Zwitserlood, P., Kooijman, V., & Hagoort, P. 2005. Anticipating upcoming words in discourse: Evidence from ERPs and reading times. *Journal of Experimental Psychology: Learning, Memory, and Cognition* 31(3). 443–467.

Van Son, R., & Pols, L. 2003. How efficient is speech? *Proceedings of the Institute of Phonetic Sciences* 25. 171–184.

Vigliocco, G., Antonini, T., & Garrett, M.F. 1997. Grammatical gender is on the tip of Italian tongues. *Psychological Science* 8(4). 314–317.

Vigliocco, G., Vinson, D.P., Paganelli, F., & Dworzynski, K. 2005. Grammatical gender effects on cognition: Implications for language learning and language use. *Journal of Experimental Psychology: General* 134. 501–520.

Wicha, N.Y.Y., Moreno, E.M., & Kutas, M. 2004. Anticipating words and their gender: An event-related brain potential study of semantic integration, gender expectancy, and gender agreement in Spanish sentence reading. *Journal of Cognitive Neuroscience* 16. 1272–1288.

Wittgenstein, L. 1953. *Philosophical investigations.* Oxford, England: Blackwell.

Wood, S.N. 2006. *Generalized additive models: an introduction with R.* CRC press.

Wood, S.N. 2011. Fast stable restricted maximum likelihood and marginal likelihood estimation of semiparametric generalized linear models. *Journal of the Royal Statistical Society: Series B (Statistical Methodology)* 73(1). 3–36.

Zipf, G.K. 1935. *The psychobiology of language*. New York: Houghton-Mifflin.

Zipf, G.K. 1949. *Human behavior and the principle of least effort: An introduction to human ecology*. Addison-Wesley: Cambridge, MA.

Zubin, D. & Köpcke, K.-M. 1986. Gender and folk taxonomy: The indexical relation between grammatical and lexical categorization. In: C. Craig (ed.): *Noun Classification and Categorization* (Typological Studies in Language; Vol. 7). Philadelphia: Benjamins, North America. 139–180.

Zubin, D. & Köpcke, K.-M. 1996. Prinzipien für die Genuszuweisung im Deutschen. In: Lang, E. and G. Zifonun (eds.): *Deutsch typologisch*. Berlin: de Gruyter. 473–491.

Zubin, D. & Köpcke, K.-M. 2009. Gender control-lexical or conceptual? In: Steinkrüger and Krifka (eds.): *Trends in Linguistics: On Inflection*. Berlin: de Gruyter. 237–262.

Index

www.ingramcontent.com/pod-product-compliance
Lightning Source LLC
LaVergne TN
LVHW010641300926
849161LV00005B/1018